# CROP-PRODUCTION IN INDIA

# CROP-PRODUCTION IN INDIA

*A Critical Survey of its Problems*

ALBERT HOWARD

GYAN

*Published by*

GYAN

Gyan Publishing House
5, Ansari Road
Daryaganj, New Delhi-110002
Phone: 011-47034999, 9811692060
E-mail: books@gyanbooks.com

*Distribution Network*
gyanbooks.com
India, USA, Canada, UK, Australia



ISBN: 978-93-7225-893-6 (PB)

First Published, 1924

2nd Impression 2025

*Printed at:* Gyan Press, Delhi.



**CROP-PRODUCTION IN INDIA**
Author: ALBERT HOWARD

# CROP-PRODUCTION
# IN INDIA

# CROP-PRODUCTION IN INDIA

## *A Critical Survey of its Problems*

by

ALBERT HOWARD, C.I.E., M.A.

DIRECTOR OF THE INSTITUTE OF PLANT INDUSTRY, INDORE, AND AGRICULTURAL ADVISER TO STATES IN CENTRAL INDIA

(FORMERLY IMPERIAL ECONOMIC BOTANIST, AGRICULTURAL RESEARCH INSTITUTE, PUSA)

1924

# PREFACE

AGRICULTURE is and for many years to come must remain India's greatest industry and the foundation of the State. In comparison with the value of the annual produce of the soil and the trade in raw materials, the remaining industries of the country are, with few exceptions, relatively unimportant.

For the last twenty years a number of men of science have been engaged in exploring the various directions in which the produce of the soil of India can be increased and improved. The results of this experiment, which was initiated by Lord Curzon in 1904, have been both interesting and encouraging. Many improvements in crop-production have been devised, which are well within the means of the people. To utilize these discoveries, an organization, working in the Districts, has been evolved by which the lessons, learnt at the Experiment Stations, have been brought rapidly and successfully to the fields of the cultivators.

Besides the results obtained on crops and the work done in the distribution of seed, the investigations in progress have brought to light a number of problems, the successful solution of which must be accomplished before further advances can be made. From their nature and magnitude, these questions can only be dealt with if the active co-operation of the public is secured. The first step in enlisting this interest is obviously to set out these problems in simple language, and to explain in what manner they affect the future progress of India. This is the chief aim of this book. It has been written with a subsidiary purpose, namely to stimulate further work and to attract a number of active and enthusiastic investigators to these questions. Both these objects will have to be attained before the immense potentialities of the soil of India can be developed.

It has not been possible, within the limits of this book, to give a complete account of the work done by the Agricultural Department in India during the last twenty years. The aim has been rather to indicate the vast field of work still untouched and, while doing so, to make use of past work as pointers in the investigations of the future.

A short, selected bibliography has been appended to each chapter so as to enable the student to explore the various subjects dealt with in greater detail.

ALBERT HOWARD.

PUSA,

*Feb.* 29, 1924.

# CONTENTS

# PART I

# THE SOIL

THE theme of the following chapters on the soil is crop-production, i. e. the life of the plant in its relation to the environment. The plant has been regarded as the centre of the subject ; the various soil factors have been considered in relation to the welfare of the crop.

In the diversity of her soils India resembles other continents. The differences between the various types are, from the point of view of geology and chemistry, both numerous and extensive. Nevertheless, from the standpoint of the plant, the soil factors which limit production are not numerous. Provided the cultivator can supply sufficient water, air, and organic matter and also maintain a suitable texture, the soils of India are exceedingly fertile and produce abundant crops. Some of these limiting factors are discussed in the first section of this book, and a number of subjects for further work have been outlined. These questions have all been dealt with from the point of view of the plant. It is suggested that some of them, such as the proper use of canal water, the utilization of the tube-well as an instrument in the development of rural India, the origin of alkali land, the best method of dealing with the nitrogen supply, and the working-out of a system of rural economy for the Canal Colonies of the Punjab and Sind, should now be taken up as definite projects. Many of these questions have already been examined, and a large amount of useful knowledge has been collected. The time has come to bring all this scattered information together, and to concentrate future effort, as far as the crop is concerned, on a few important problems.

# I

# SURFACE-DRAINAGE AND EROSION

In a country like India, which is liable to long periods of drought, the first condition of success in the improvement of agriculture is the regulation of the rainfall after it reaches the soil. We cannot control either the amount or the irregular distribution of the monsoon, but a great deal can be done in regulating it for the benefit of crops after it has fallen. The necessity for the conservation of water is at once apparent from the short period—June to September—during which most of the rainfall of the South-West monsoon is received and from the fact that a constant supply of moisture is needed by crops for a large portion of the year.

At first sight, the most obvious direction in which the water-supply can be controlled for the benefit of agriculture is by means of irrigation from rivers whereby the surplus drainage can be spread over the country by a network of canals. A little consideration, however, will show that this idea is incomplete. Of equal value to canal-irrigation is the regulation of the run-off or, in other words, the provision of surface-drainage. The regulation of the surface-drainage should precede irrigation, as it is clearly of greater advantage to make the best use of the rainfall, which costs nothing, before going to the expense of leading river-water to the fields for the purpose of making up any shortage of moisture. In considering the general question of the regulation of the rainfall, both by surface-drainage and also by irrigation, it is well to bear in mind how fundamentally crops differ among themselves in their need of moisture. The methods adopted must conform to these requirements. Dry crops like maize, millets, cotton, wheat, and pulses require a comparatively small quantity of water, which, moreover, exercises its maximum effect only when all due attention is paid to the physical condition of the soil. On the other hand, rice, and to a less extent jute, are examples of water-culture. In these cases the volume of water needed for growth is very great. Further, the jute crop requires a large supply of constantly fresh water at a suitable temperature for retting and for the cheap transport of the fibre. Control of the run-off may, therefore, vary with the crop grown. How important these facts are will be evident when we consider the regulation of the water-supply in the delta of the Ganges.

It is during heavy falls of rain that control of the drainage is most necessary in India. These are so violent that a large portion of the water runs off the surface towards the drainage lines, carrying with it the most valuable portion of the soil—the fine particles and a large part of the organic matter. Sometimes the drainage from the higher land causes water-logging on its way to the swamps and rivers. Both on the alluvium of the plains and on the soils of the Peninsula, there is a very general absence of regulation of the surface-drainage. The surplus water rapidly runs to waste, and often there is no time for it to soak into the soil. Constant erosion, with consequent loss of fertility, accompanies this waste of valuable water. Sometimes there is insufficient water in the subsoil to tide the crops over a break in the rains, and the permanent fertility of the fields is lowered by the loss of its most valuable constituents. In other cases the soil has to deal with both rainfall and run-off, and water-logging results. These facts help to explain why a well-distributed rainfall, even when below the average, is so important in India. When the showers are light and frequent there is ample time for absorption without water-logging, while at the same time the loss of fertile silt by erosion is very small.

Examples of the evil consequences which result from want of control of the surface-drainage are unfortunately only too abundant. Thousands of acres of valuable land on the left bank of the Jumna have been destroyed by the formation of a network of ravines which produce little more than a crop of grass in the rains. These ravines have been carved out of the soft alluvial soil by the uncontrolled drainage in the past. Every year they extend farther and farther from the river until, at the present time, they measure many hundreds of yards in length. Villages, which at one time were surrounded by fertile fields, now lie in a network of useless gullies. It is true that successful experiments in the afforestation of this strip of desert land are being undertaken by the Forestry Department and that in-time a supply of useful timber and better fodder will result, but the area devastated is far too large to be rapidly reclaimed in this way. Further, the expense will be considerable. The real remedy for such damage is prevention—the control of the drainage in the first instance. In matters such as this, little can be hoped for from individual cultivators, as they are too intent on their small areas of land and too poor to execute a drainage scheme for the country-side.

Less striking than the ravine lands of the Jumna, but far more extensive and therefore more important, is the erosion which goes on on the soils of the Peninsula—in Central India, Gwalior, the Central Provinces, and Bombay. Some eighty years ago Sleeman drew pointed attention to the damage done by uncontrolled drainage in these areas in the following words: 'I am disposed to think that the most productive parts of the surface of Bundelkhand, like that of some of the districts of the Nerbudda territories, which repose on the back of the sandstone of the Vindhya chain, are fast flowing off to the sea through the great rivers which seem by degrees to extend the channels of their tributary stream into every man's field, to drain away its substance by degrees, for the benefit of those who may in some future age occupy the islands of their delta. I have often seen a valuable estate reduced in value to almost nothing in a few years by some new antennae, if I may so call them, thrown out from the tributary streams of great rivers into their richest and deepest soils. Declivities are formed, the soil gets nothing from the cultivator but the mechanical aid of the plough, and the more its surface is ploughed and cross-ploughed the more of its substance is washed away towards the Bay of Bengal in the Ganges, or the Gulf of Cambay in the Nerbudda. In the districts of the Nerbudda we often see these black hornblende mortars, in which sugar-canes were once pressed by a happy peasantry, now standing upon a bare and barren surface of sandstone rock, twenty feet above the present surface of the culturable lands of the country.' Sleeman's remarks are still true to-day except in those cases where enlightened administration has encouraged and assisted the people to check this denudation by means of embankments. Nothing strikes the traveller, during the rains in the black soil areas of the Peninsula, more than the universal scouring of the fields by the run-off and the enormous annual loss of the best portion of the soil. If only the surface-drainage were controlled, this loss of fertile soil would stop and time would be given for the water to soak into the soil. This increased absorption would check erosion, would lead to better crops, and would also raise the spring-level and thus maintain the wells in action during the cold season and the succeeding hot weather. In some areas the soil of whole valleys has been removed by denudation, and the rocky subsoil left only maintains with difficulty a thin covering of scrub. Soil-formation, however, is going on even in such tracts, and it is extraordinary how quickly fertile land can be re-created by means of properly constructed embankments stretching across the valley. In the Gwalior State examples of such reclamation are numerous, and fine stretches of wheat are now being grown on the soil held by these embankments. In Bombay many other examples of the successful control of rain water, after it has fallen, exist. These not only indicate the remedy for a state of things which leads to a great annual drain of the natural capital of India, but also prove how rapid is the decay of the rocks and how much new soil is being created every year. Although erosion is extensive, it is partly counterbalanced by the formation of fresh soil. The position, therefore, is not hopeless provided denudation can be stopped and the yearly accretions of new earth can be collected and retained.

It is in the planting areas of the East, however, that the most striking examples of soil denudation are to be found. Instances of damage to the natural capital of the country are to be seen on the tea estates near Darjeeling, on the hill-sides in Sikkim, on the upper terraces in the vale of Kashmir, in the Kumaon hills, on the tea estates in Ceylon and Assam, and in the planting districts of Southern India and the Federated Malay States. In most of these areas forest land was so abundant that the need for the preservation of the soil was not at first recognized. Thanks to the efforts of Hope, a former scientific officer employed by the tea industry in Assam, the control of the drainage and the checking of erosion are now widely recognized and are being dealt with by the planters in many parts of India. A great impetus to this work was given by the publication in India of a detailed account of the methods in use by the Dutch planters in Java, where the terracing and drainage of sloping land under tea and other crops has been carried to a high state of perfection. In this island the area of land available for planting is strictly limited, while the feeding of the large indigenous population is always a serious problem. As a consequence the development of the island is very strictly controlled by the Government, and one of the conditions of planting new forest lands is the provision of a suitable system of terraces, combined with surface-drainage. The advantage is not all on the side of the State. The manuring of tea soils in Java is far less necessary than in Ceylon and India, while one important consequence of the retention of the valuable soil made by the forest is healthy growth which suffers remarkably little damage from insect and fungoid pests.

The remedy in cases such as these is simple. A system of embankments, provided with spill-ways, is all that is needed in the worst cases of denudation. On sloping hill-sides, terracing combined with surface-drains on the Java system, or embanked fields such as are met with in Bombay and the Central Provinces, are essential. Where the slope is small, shallow surface-drains, provided with slightly raised grass-borders, on the lines of a system worked out at Pusa, is sufficient to check the run-off, to assist absorption, and decrease erosion. The surplus water, instead of water-logging the areas passed over, can then be led into low-lying rice areas or into the streams or rivers. The great thing in such cases is to follow the despot's maxim—to divide and rule—and so prevent a rush of water over the surface towards the drainage lines, which is certain to carry with it a large portion of the most fertile soil and to produce water-logging.

There are other directions in which surface-drainage can be controlled besides the checking of erosion and the prevention of water-logging. A great deal can be done for crops like rice and jute which are largely grown in water-culture. Instances of such control are naturally to be looked for in the deltas of the great rivers which discharge into the Bay of Bengal—the Ganges, the Godavery, and the Kistna. In the case of the Godavery and the Kistna, rice cultivation has benefited greatly by the anicuts thrown across these rivers at the head of the delta, by means of which the water running to waste has been carried to the rice fields by a network of canals. India owes these highly successful examples of the regulation of the surplus drainage to the genius of the late Sir Arthur Cotton. When, however, we turn to Bengal, where the conditions are quite different, the early efforts in control took a different form, namely the embankment of the rivers. Much light has been thrown on these matters by the investigations of Bentley on malaria and of Addams-Williams into the principles underlying successful drainage under deltaic conditions. Bentley found in the delta of the Ganges that an intimate relation exists between the prevalence of malaria and the inundation of the rice areas. In tracts such as East Bengal, which are subject to annual inundation, the rice crops are excellent, there are practically no waste lands, the population is large, and malaria exists to so slight an extent as to be almost negligible. In parts of Central and West Bengal, however, quite a different state of things exists. The greater portion of the Districts of Nadia, Murshidabad, Hooghly, and Burdwan have been deprived of the natural inundations to which they were once subject by the construction of river-embankments. In the old days, sixty or seventy years ago, these embankments were kept in such bad repair that they did not prevent the flooding of the country, and their harmful effect was not apparent. When they were made really efficient barriers to the inundation, the protected areas began to suffer in various ways. The fertility of the soil diminished, the area of waste land increased, population declined, and with it malaria increased to an appalling extent. The independent investigations of Addams-Williams on the drainage of deltaic Bengal have led to very similar conclusions. This investigator has stated very clearly the principles which should be followed in any attempts to regulate the drainage of areas like the Gangetic delta. The first condition is to prolong the natural life of the rivers which carry off the silt-laden water. The second is to postpone the construction of embankments till the level of the land has been raised by silt deposits as much as possible. Premature river-embankments and any artificial deflection of the general drainage by roads and by railway-embankments lead to the rapid silting up of the rivers and to the destruction of nature's drains. The general blocking of surface-drainage follows, and the water-logging of the country and a diminished rice crop are the inevitable results. It is fortunate that in addition to the results of Bentley on the incidence of malaria and of Addams-Williams on the life-history of a Bengal river, definite scientific evidence of the requirements of the rice plant is also available. The researches of Harrison and of his staff at Coimbatore in Madras on the gases of swamp rice soils have drawn particular attention to the importance of the oxygen supply of the roots of the rice crop by means of a slow movement of aerated water through the upper layers of mud in which this crop grows. The roots of rice must have a constant supply of oxygen. As they are immersed in mud and water, the only way this substance can be provided is in solution in water which must move slowly through the soil. These results help to explain the good effects which follow the successive inundations of river water over the rice areas in Bengal. These slowly change the water and renew the supply of dissolved oxygen for the roots of the rice. Taking the results of these investigators together, it will be clear that as far as the prosperity of Bengal is concerned the rice plant is the centre of the subject, and that from the essential needs of this crop we can deduce the policy which should be followed in the general development of the country and in the prevention of malaria. If we control the water-supply for the benefit of rice and jute, everything else will follow. The rivers will tend to keep themselves open, the country will produce food and fibre in abundance, and the inroads of malaria will begin to decrease of themselves. While communications in the shape of railways and roads are essential in the modern state, nevertheless their construction must conform to the natural drainage of the country and there must be no interference with agriculture.

The larger aspects of this question must now be considered.

So many different authorities, for the most part independent of each other, are concerned with the general surface-drainage of the country, that unless some guiding principle is recognized and acted upon, chaos is inevitable. The simplest and most effective remedy appears to be the construction of drainage maps on the system devised by the late Sir Edward Buck in 1870 when Settlement Officer of the Farrukhabad District. This consists in marking on an ordinary map the direction in which rain-water runs off the land. By this means the drainage lines can be determined easily and cheaply. Once drainage maps, preferably drawn up by competent engineers who possess the necessary agricultural insight, are available, it will be easy to control all such undertakings as the construction of roads, railways, canals, and embankments, and to see that nothing interferes with crop-production. From the drainage map it is a short step to the preparation of a series of monographs on the river-systems of India such as already exist in European countries like Italy. One of the most suitable areas for such projects is the submontane tract of the United Provinces which lies between the Gandak and the Gogra and comprises the Districts of Gorakhpur, Basti, Gonda, and Bahraich. A drainage survey of this area is strongly recommended in the recent *Report of the Sugar Committee*, and its successful execution would constitute an important landmark in the development of agriculture—India's greatest industry.

## BIBLIOGRAPHY

Addams-Williams, C.—Drainage problems in the Ganges delta, Calcutta, 1913.

Anstead, R. D.—The prevention of soil erosion on tea estates in Southern India, *Agr. Jour. of India*, xiv, 1919, p. 787.

Benskin, E.—Afforestation in the United Provinces, Allahabad, 1921.

Gilbert, T.—Field embankments, *Agr. Jour. of India*, xvi, 1921, p. 483.

Harrison, W. H., and Subramania Aiyer, P. A.—The gases of swamp rice soils, *Mem. of the Dept. of Agr. in India* (*Chemical Series*), iii, 1913, p. 65, and iv, 1914, p. 1.

Hope, G. D.—Note on soil denudation by rainfall and drainage: conservation of soil moisture, *Agr. Jour. of India*, xi, 1916, p. 134.

Hope, Lady.—Life of Sir Arthur Cotton, K.C.S.I., London, 1900.

Howard, A.—Soil erosion and surface-drainage, *Bull. 53, Agr. Research Institute, Pusa*, 1916.

Howard, A., and Howard, G. L. C.—Drainage and crop-production in India, *Agr. Jour. of India*, xiv, 1919, p. 377.

Marsh, H.—Protective works in Central India, *Agr. Jour. of India*, i, 1906, p. 290.

Proceedings of the Third All-India Sanitary Conference, 1914.—Supplement to the *Indian Jour. of Med. Research*, 1914.

Proceedings of the Board of Agriculture in India, Calcutta, 1916.

Report of the Indian Sugar Committee, Simla, 1921.

Sleeman, W. H.—Rambles and recollections of an Indian Official, Oxford, 1915, p. 209.

Smythies, E. A.—Afforestation of ravine lands in the Etawah District, United Provinces, *Indian Forest Records*, vii, 1919, p. 217.

Stockdale, F. A.—Soil erosion, *The Tropical Agriculturist*, lxi, 1923, p. 131.

# II

# SOIL-AERATION

THE usefulness of the river-systems of India is not confined to the removal of surplus water. They also help to ventilate the soil, and in particular to increase its supply of oxygen and nitrogen from the atmosphere. During the rains the level of permanent water in the subsoil is at its highest point, after which the rivers exert a continuous pull and, as it were, slowly drag water out of the earth. The fall in the level of the ground-water is accompanied by an in-draught of air into the soil from the atmosphere. The rivers and streams act, therefore, on the soil like a subsoil drain. How important this aerating action of the rivers to agriculture is can be seen in tracts like North Bihar during periodical floods, when the rivers are unable to cope with the surplus water and with the heavy falls which rapidly follow one another. Soil-aeration stops as percolation is retarded by the water-logged surface soil. After a time there is a shortage of air in the pore-spaces. The flora and the chemistry of the soil are then profoundly affected. The crops suffer and the yield falls off, the amount of damage closely corresponding with the degree of interference with the air-supply.

The full significance of soil-aeration in agriculture is only now being recognized. Till recent years the agricultural Experiment Stations have been situated in humid regions where the rainfall is well distributed. Rain is practically a saturated solution of oxygen, and is very effective in supplying this gas to the soil whenever percolation is possible. Hence in such regions, crops are not likely to suffer from poor soil-aeration to anything like the same extent as those grown in the arid regions of North-West India where the soils are silt-like and lose their porosity when the surface is flooded, followed by the formation of an impermeable surface crust.

The aeration of the soil is best regarded from the point of view of ventilation. It closely corresponds with changing the air in a living room. As is well known, the soil is not solid but consists of particles with intervening spaces between them, known collectively as the pore-space. One of the objects of cultivation is to increase the area of the pore-space or, in other words, to improve the soil texture. The pore-space is taken up by two things—air and water. The water occurs in thin films round the soil particles while the soil air fills up the rest of the pore-space. In these water films there is intense biological activity. This is of two kinds. In the first place, the roots of crops (by means of the fine root hairs) are constantly engaged in the absorption of the water and food materials essential for growth. This work involves incessant protoplasmic activity in which oxygen is used up and carbon dioxide produced as a by-product. The work of the roots is, therefore, a respiratory process. For this to proceed, continuous supplies of oxygen must pass into the soil from the atmosphere, and at the same time the excess carbon dioxide must diffuse out in the reverse direction. The second group of activities in these water films is brought about by the various soil bacteria occupied in the decomposition of organic matter. These organisms (like the roots of crops) are alive and are constantly respiring. They compete with the roots for oxygen and also produce carbon dioxide in large quantities. Everything goes well in this breaking down of the organic matter by means of the soil organisms if there is sufficient ventilation. Harm at once results, however, if the supply of oxygen falls off and if the fermentation is allowed to go on too long in the absence of air. Another group of organisms then become prominent which are able to obtain their oxygen from various substances containing this element in combination. Intermediate products of decay, toxic to the roots of plants, are formed, and the valuable combined nitrogen is set free as gas and lost to the soil. The result of poor soil ventilation is, therefore, a greatly reduced crop.

Interference with aeration during the monsoon soon produces very definite effects. This is proved by the result of experiments at Pusa, of which one example may be given. In 1910 the artificial water-logging of wheat land during the month of September reduced the subsequent yield by nearly twelve maunds (984 lb.) of grain to the acre. The loss of crop was proved to be due to the destruction of combined nitrogen which resulted from the water-logging and the cutting-off of the supply of air to the soil. Over a very large part of the plains of India similar damage takes place every monsoon. In most fields there are slight concavities into which the surplus rainfall collects. Incipient water-logging follows. The valuable nitrates are destroyed and the texture of the soil deteriorates. On these partially water-logged areas the succeeding cold season crops yield at least 50 per cent. less than the best portions of the fields. As these patches often cover as much as a quarter of the area and as they are almost universal, the destruction of combined nitrogen in the plains every year must run into thousands of tons. The annual value of the loss of crop must amount to crores of rupees. The remedy is simple and within the means of every cultivator. The surface of each field should be graded so that absorption is uniform. The surplus water should be got rid of by surface drains, and not allowed to run over other land on its way to the drainage lines.

The soil has other uses for air beyond the oxidation processes. The nodules on the roots of leguminous crops work up gaseous nitrogen into food materials which can be used by the plant. Obviously, this can only go on if soil-aeration is adequate. The chief rôle of these root-nodules is the fixation of free nitrogen, and the process is a definite gain to agriculture from the point of view of the provision of plant food. Nitrogen fixation, however, is not confined to the root-nodules of leguminous plants. It also takes place in tropical soils, provided they are aerated and provided moisture and organic matter are both present. The records of the past prove clearly how important nitrogen fixation is in Indian soils. In Rohilkhand for example, sugar-cane has been grown for centuries without manure, and the yield does not appear to have undergone any diminution. It is about the same at the present time as that given in the *Ain-i-Akbari*. This result is only possible through nitrogen fixation by *Azotobacter* and other soil organisms, a process which requires constant aeration and is, therefore, stimulated by the numerous ploughings which are given during the fallow preceding the cane crop. Work on this subject is in progress at several centres in India with the object of discovering the optimum conditions for fixation. Once these are known, it may be possible by simple means to increase the nitrogen supply for the crop and to bring about increased fertility at a small cost.

Perhaps the most interesting evidence so far obtained on the importance of soil-aeration has been furnished by the plant itself. In areas of the plains such as North Bihar, where aeration is difficult, we should expect that the root-system of indigenous crops would be as near as possible to the surface. This is so. On the other hand, on the black soils of the Peninsula, where the land cracks extensively during the cold weather and dries out completely near the surface, we should expect the general root-development to be deep, as the moisture is well below the surface and the extensive cracking promotes aeration. The varieties of linseed and opium poppy grown on the black soils are all deep rooted compared with the types met with on the alluvium.

A large amount of very definite experimental evidence is available in India on the importance of soil-aeration. At the Forestry Research Institute at Dehra Dun, Hole has completed a study of the factors involved in the natural regeneration of *sal* (*Shorea robusta* L.) forests. One of the most important needs of the young seedlings is adequate soil-ventilation, without which their establishment is slow and uncertain. The same factor appears to be equally necessary in established forests. Cases have been examined where, on account of poor soil-aeration, *Shorea robusta* practically ceases to grow during the rains, which ought to be the period of most vigorous development. The rapid regeneration of the *sal* forests of India is a matter of great importance both from the point of view of the timber supply and also of the revenue.

Another interesting line of investigation on soil-aeration relates to the increase of the air-supply of close soils by the addition of inert materials like sand or coarse brick dust. These open up the soil and improve the aeration. In experiments at Pusa this treatment has resulted in an increase in the yield of from 10 to 40 per cent. according to the crop. Such experiments of course are not directly applicable to the fields of the cultivators, but they are of value in proving the need of more air. Their publication helped to draw attention to the value of permeability and to the great possibilities of soils with good natural drainage. If aeration is such an important growth factor, it would follow that soils possessing great permeability would often prove to be of the greatest value. Clouston, working in the Central Provinces, was the first to make any practical use of these ideas in India. In this Province large areas of poor laterite soils occur which were considered below the margin of cultivation. They produce, for the most part, a thin growth of grass in the rains with an occasional millet crop. Although exceedingly poor in the ordinary sense, these soils possess excellent porosity and a texture which is unaffected by irrigation. By the addition of organic matter and by providing irrigation facilities, these waste areas can be transformed into garden land of the highest quality. Crops of over forty tons of stripped cane to the acre have been grown at Chandkhuri near Raipur, as well as heavy yields of cotton, ground-nuts, indigo seed, and fodder. Besides these poor laterite soils, attention has also been paid to the development of the sandy tracts in the Eastern Punjab by means of water raised by tube-wells from the subsoil. Here again an open texture and aeration above the average have proved to be valuable assets. In both these cases the significance of the soil-aeration factor has directed attention to the possibility of opening up areas which till recently were considered to be useless and were classed as waste lands. We now know that these tracts constitute a valuable asset, and that their development depends only on a supply of irrigation water and of organic matter.

Many of the soils met with in India are not well aerated naturally. The proper tillage of these soils is important, and they present a number of problems which need investigation. So far in the history of agriculture the methods of cultivation and the design of implements have been very largely the result of trial and error. Frequently no guiding principle beyond experience has been recognized. The value of growth factors like aeration, coupled with the main facts of root-development, should help to raise the subject of tillage to a higher plane and lead to improvements in cultivation and to the design of more efficient implements. The cultivator in the plains is at present handicapped in the proper management of monsoon fallows by the want of a suitable instrument. He understands, from long experience, the beneficial effect of constantly stirring the surface soil during the rains, but with the country plough it is impossible to get over the land quickly enough. He needs an implement which will multiply threefold the efficiency of his cattle. The five-tine spring-tooth cultivator was designed for this purpose just before the war, and has been on the market in India for some time. In some parts of the plains its use is understood, and the cultivators are beginning to adopt it. As it spreads among the people, the area under wheat and sugar-cane will tend to increase, as one of the difficulties in growing these crops is the proper management of the preceding monsoon fallow with the cattle power available.

Another group of problems centres round the maintenance of the permeability of the soil during the rains. Both the black soils and the Gangetic alluvium often refuse to drain after long continued wetting, and pass into a condition which prevents aeration altogether. The result is associated with the formation of colloids, but up to the present little attention has been paid to it by investigators. If simple means of maintaining percolation, such as the addition of small quantities of sulphur, could be devised, a great step forward would be possible. On the black soils, subsoil drainage has been shown by Allan to be of great value in maintaining soil-aeration during the monsoon and in increasing production. This method of assisting the percolation is, however, impracticable in many parts of the alluvium, as the drains rapidly become clogged with silt.

Another very interesting aeration problem is bound up with the quality of plant products, which, as is well known, varies greatly with the locality. What are the factors on which quality depends? The breed or variety is certainly one. A rough, short-stapled cotton, for example, can never be transformed by alteration in the environment so as to resemble the best Egyptian or Sea Island types. Such a cotton can be improved to a limited extent only, but even under the best circumstances the fibres will always remain coarse and short. Besides the variety, the environment undoubtedly has an influence on quality. What is the nature of the factor involved in the environment? It is suggested that the factor in question is soil-aeration. Several Indian examples can be quoted in support of this view. In the case of tobacco, certain tracts, such as the Perganah Saraisa in Tirhut, Jais in the District of Rai Bareli, and the Mustung valley in Baluchistan, have achieved a reputation for quality which is well known in the trade. In all these places, the soils which produce the best qualities are those in which aeration is above the average. In the Central Provinces, Clouston has shown that the same variety of cotton grown on porous laterite soils has a better staple than when grown on heavy black soils side by side in which the aeration is poor. In the United Provinces the sugar produced on the open soils of the Meerut Division is of a better quality than that from the stiffer areas of the Gorakhpur District. These matters are certain to repay investigation. If we knew what the essential conditions for the full development of quality are, the knowledge might easily lead to useful practical developments.

A number of workers are devoting attention to soil-aeration problems both in India and other countries. In the United States particularly is this the case. The investigations of Cannon in Arizona, of Clements at the Carnegie Institution of Washington, and of Livingston at the Johns Hopkins University, are throwing considerable light on the factors involved in soil-aeration. This matter is also receiving attention at the Rothamstead and Long Ashton Experiment Stations in Great Britain. At any moment these labours might open up further possibilities in the agricultural development of India.

## BIBLIOGRAPHY

Allan, R. G.—Soil drainage, *Bull. 85, Agr. Research Institute, Pusa*, 1919.

Cannon, W. A., and Free, E. E.—The ecological significance of soil-aeration, *Science*, xlv, n.s., 1917, p. 178.

Clarke, G.—Notes on improved methods of cane cultivation, Allahabad, 1919.

Clarke, G., Banerjee, S. C., Naib Husain, M., and Qayum, A.—Nitrate fluctuation in the Gangetic alluvium and some aspects of the nitrogen problem in India, *Agr. Jour. of India*, xvii, 1922, p. 463.

Clements, F. E.—Aeration and Air-content: the rôle of oxygen in root activity, *Publication No. 315, Carnegie Institution of Washington*, 1921.

Clouston, D., and Padmanabha Aiyer, A. R.—The physical texture of soils in its relation to crop-production, *Agr. Jour. of India* (*Indian Science Congress Number*), xiii, 1918, p. 89.

Hole, R. S.—Ecology of sal (*Shorea robusta*), *Indian Forest Records*, v, 1914–18.

Howard, A.—Soil aeration in agriculture, *Bull. 61, Agr. Research Institute, Pusa*, 1916.

————The improvement of crop-production in India, *Jour. of the Roy. Soc. of Arts*, lxviii, 1920, pp. 555 and 569.

Howard, A., and Howard, G. L. C.—Soil ventilation, *Bull. 52, Agr. Research Institute, Pusa*, 1915.

————The economic significance of the root-development of agricultural crops, *Agr. Jour. of India* (*Indian Science Congress Number*), xii, 1917, p. 17.

————Some aspects of the indigo industry in Bihar *Mem. of the Dept. of Agr. in India* (*Botanical Series*), xi, 1920, p. 1.

————Note on the theory of phosphatic depletion in the soils of Bihar, *Agr. Jour. of India*, xviii, 1923, p. 148.

Howard, A., and Hole, R. S.—Recent investigations on soil-aeration with special reference to Agriculture and Forestry, *Agr. Jour. of India*, xiii, 1918, p. 416.

Mukerjee, Jatindranath.—Carbon dioxide in soil gases, *Agr. Jour. of India*, xix, 1924, p. 146.

Sethi, D. R.—Reclamation of the desert area of the Kapurthala State, *Agr. Jour. of India*, xvii, 1922, p. 546.

## III.

## IRRIGATION AND WATER-SAVING

In addition to the regulation of the run-off, the utilization of the surplus water, by means of irrigation, is of the greatest importance in the development of Indian agriculture. Considerable progress has been made in this work, not only by means of perennial canals but also by the utilization of subsoil water by the tube-well.

Irrigation can be regarded from two points of view—that of the engineer, whose function it is to deliver periodically a certain volume of water at various points, and that of the plant which has to use this water for growth. The first is the engineering, the second the agricultural aspect. It is unnecessary to devote any attention to the engineering side of the subject, as this matter is already being developed as rapidly as the resources of the State permit. As every one knows, a large number of canals are already in operation, while vast projects, such as the Sukkur barrage in Sind, the Sarda Canal in Oudh, and the Sutlej scheme in the Punjab and Rajputana, are being set in motion. The agricultural side of the subject, however, is not so satisfactory. It is with this aspect of the question that it is now proposed to deal. Stated in general terms, the problem is a two-fold one—the discovery of the best use of the water provided and of the easiest method of increasing the revenue. Both these matters are important. It is of little advantage to India for the engineers to provide water if this gift is misused and the natural fertility of the country-side suffers. The aftermath of the war and the institution of popular government both point to the necessity for an increase in the revenue wherever this is possible.

The first thing that strikes the traveller in the canal-irrigated tracts of Northern India is the waste of water which goes on on all sides. The small channels of the cultivators, which lead the water to the fields, are so badly kept that bursts are frequent. The grading of the surface of the fields is poor. The field compartments (*kiaris*) are too large and often do not exist at all. The land is frequently unevenly watered and the subsequent cultivation is not uniform, leading to a poor tilth. There is a great contrast between the application of well-water in the Eastern Districts of the Punjab and irrigation in the Canal Colonies. Where the cultivator has to lift the water from wells, the work is a pattern of neatness and order. Where canal-water is used, untidiness and waste are the rule. The difference appears to be due entirely to the system. In the first case, the cultivator pays for the water he uses by the construction of a well and by the provision of work cattle to raise it. In the second case, the water is provided for him and he is assessed not by the amount used but by the area irrigated. So the waste of water is not his business. In addition to the loss of water, actual damage is often done to the locality. Round Amritsar, for example, the country became so water-logged that the surplus water has had to be lifted from the soil by means of tube-wells of a special pattern. On the Nira Left Bank Canal in the Deccan, so much water has been applied that of the total commanded area of 81,000 acres, 9,100 acres have been transformed into useless alkali land, while the area appreciably damaged is estimated at 27,000 acres. Reclamation is only possible by an expensive drainage system. Extreme examples such as these are fortunately the exception rather than the rule. Minor injuries to the crops are, however, very common on most of the perennial canals. Growth is hardly ever quite normal. In the case of wheat, for example, the crop dries up rather than ripens, the process is delayed, and the grain is not really well filled. Canal-irrigation in the hands of the cultivator seems to put a brake on the wheel of life.

The amount of excess water applied to crops like wheat in Upper India appears to be from 30 to 50 per cent. This is indicated by the result of recent experiments. The first investigations were designed to ascertain what kind of wheat crop could be grown without any watering at all after sowing. These were first carried out at Quetta, and the land was watered once only, just before sowing. The average yield of grain was nearly eighteen maunds[1] to the acre. The next step was to compare the crop grown on one watering after sowing with that

[1] One maund = 82 lb.

obtained when three waterings were given. Increasing the number of waterings *decreased* the yield by 26 per cent., a result probably due to the thick impermeable surface crust interfering with the aeration of the soil. Similar results were then obtained in the Punjab, in Sind, and in the United Provinces. In the Punjab in 1916–17 an average of nearly ten maunds of grain to the acre was obtained by cultivators on the watering before sowing. An extra irrigation raised the yield to sixteen maunds, while two waterings after sowing lowered the yield by a maund to the acre. At Mirpurkhas in Sind the average yield obtained on the irrigation before sowing was eighteen maunds to the acre, and no increase took place when a second watering was given. At the Sugar Experiment Station at Shahjahanpur a most interesting result was obtained in 1919. On land in good condition after sugar-cane, and on a field over three acres in area, a yield of thirty-six and a half maunds of wheat to the acre was obtained when the crop was sown on the natural moisture and received afterwards only one watering. A closely similar result was obtained the same year at Quetta, where a crop of thirty-two maunds of wheat was obtained on an acre plot after Persian clover on the watering before sowing, without any subsequent irrigation. That heavy yields are possible with the minimum of irrigation water when applied to land in good condition is in accordance with theory. The object of scientific irrigation is to stretch the water thinly over the internal surface of the pore-spaces, which is many times greater than the area watered. This internal surface of the soil is greatest when the texture has been improved by the addition of organic matter and by good cultivation. Hence the better the soil texture, the smaller the volume of irrigation water needed, provided of course all the operations are skilfully conducted. The duty of water, therefore, varies according to the texture of the soil and also to the way it is managed. When it is remembered that from three to four heavy irrigations are the rule for wheat in Upper India, it will be evident from these experimental results what a vast volume of water is annually poured on to the land to no purpose. The cultivators constantly clamour for more water, as they have found by experience that abundant irrigation makes up to some extent for poor cultivation. Soils in poor condition need more water than those in good tilth. If the cultivators paid for water by volume they would learn to improve their methods and use less. Cheap and abundant water is a doubtful blessing if the matter is looked at from a broad standpoint.

The possibilities of water-saving on the black soils are being investigated at Manjri near Poona. By an alteration in the method of cultivation of sugar-cane and of applying the water, an appreciable saving has been obtained under Experiment Station conditions, and it is not unlikely that further reductions will be found to be possible. The next step is to discover how far these methods can be utilized by the cultivators. The Manjri results are of great interest, as they have been obtained in a tract where the over-watering and over-manuring of the cane crop by cultivators, above the average in intelligence and energy, have reached the point of absurdity. The cane in parts of the Deccan is really grown in water-culture somewhat like rice, and disaster is averted by the great natural permeability of the subsoil.

The time has come when the whole subject of the proper use of canal-water in India should be looked at with a fresh eye. If the present methods involved only a waste of water, the matter would only be important from the point of view of the loss of possible revenue. There seems no doubt that the damage does not end here, and that over-irrigation is slowly lowering the producing power of the land. More work is needed on the water requirements of the various crops and on the methods of agriculture best adapted to irrigated conditions. The most suitable system of rural economy, based on canal water, needs to be worked out in Sind, in the Punjab, and in parts of the United Provinces. There is no doubt that a large volume of water can be saved on the wheat and gram crops. This can be utilized for the growth of fodders like lucerne and berseem, the areas for which should receive most of the available manure. These, besides helping to solve the fodder problem, would supply useful rotation crops, and would also improve the fertility of the land. The cultivation of gram on the preliminary watering before sowing would enable the soil to rest from irrigation for a whole year before the next wheat crop. In this way the well-known beneficial effects of the Sind fallow could be easily introduced into Punjab agriculture. The first step seems to be to work out, on a practical scale, an improved system of agriculture in the Canal Colonies, based on a combination of water-saving, fodder-production, and better methods of cultivation. This would provide reliable data for a critical examination of the manner in which canal-water is used at the present time. In addition to this work, the possibility of the sale of irrigation water by volume to large cultivators, co-operative irrigation societies, and to large villages needs to be explored. This does not appear to be impossible when it is borne in mind how the primitive tribes in Baluchistan distribute the water of the *karez* (a perennial canal in miniature) according to a time-table, without any intervention on the part of the subordinates of the Irrigation Department. The result of these investigations, if adequately carried out, would in all probability lead to a change in the system by which canal-water is charged for. The present method of assessment, according to the area watered and according to the crop grown, would gradually give place to the sale of water by volume as suggested by the last Irrigation Commission. A great incentive to water-saving would then be automatically provided. It should not be impossible to convince the cultivator that the correct use of water is to his advantage, and that he must look at the subject not only from the point of view of the current crop, but also from that of the permanent fertility of his holding. A great saving of water is bound to result, after which the existing supplies would be able to command a larger area. This would lead to a gradual increase in the revenue.

In recent years the engineers in India have developed a form of irrigation intermediate between the perennial canal and the ordinary well. This is the strainer tube-well, a device by which the water present in deep-seated layers of sand can be raised to the surface by a pump driven by an oil-engine. These installations are often 250 feet in depth, and are capable of commanding from 200 to 400 acres. The system is really a canal in miniature, and is proving to be a very suitable form of irrigation for permeable soils such as are often met with in the United Provinces. In such areas perennial canals are unsuitable, as the water lost by percolation would mean a great loss of revenue and would also soon water-log the country-side. Besides being an instrument in development, the large tube-well is likely to be of great use for experiments in irrigation. The amount of water raised, the cost of lifting it, the area irrigated, the loss in transit to the fields, the value of the crops grown, as well as the gradual improvement of the land, are all matters easily determined. This information would enable the economics of this form of irrigation to be worked out in detail, and would also throw light on the question of what is a fair rate for irrigation water generally and how much is needed by the various crops. Existing canals have to be conducted according to a set of rules, and are not very suitable for the kind of investigation needed. Something more elastic is desirable, and this is possible with these new wells. Besides its value in working out the economics of the subject, this form of irrigation is well adapted for experiments in distribution. Tube-wells, operated by co-operative irrigation societies, would bring to light the difficulties which have to be overcome in the distribution of a known volume of water to individual cultivators. If directed with intelligence, bulk sale by volume should stimulate every kind of water-saving device, both in the channels and in the fields. Furrow-irrigation is an important method of water-saving, and could be adopted in the growth of fodder crops, tobacco, and vegetables. It possesses great possibilities in extracting the highest duty of water.

Little has been done in utilizing the tube-well as a weapon with which to attack the larger problems of rural development. A scheme is in contemplation in the Punjab by which the water-power of the Sutlej can be transformed into electrical energy and carried to centres like Amritsar, Lahore, and Delhi. Some of this current might be usefully employed to drive a series of tube-wells somewhat after the method already in use at Amritsar for drainage purposes. In this way the cultivators, in well-irrigated tracts like the Jullunder District, might find it an advantage to join forces and obtain their water from a series of large tube-wells rather than from a multitude of small ones. The power exists in the hills, the supply of water in the subsoil is inexhaustible, and the modern tube-well provides the means. Experiments are needed by which the economics of this question can be set out in detail. If, as appears to be the case, electric energy can be utilized in raising water for irrigation, a useful outlet for the Sutlej current will be ready to hand. One practical difficulty, however, has still to be overcome, namely the gradual blocking of the strainers by the deposition of impure calcium carbonate. This leads to a great loss of efficiency. Some method of lifting the strainers for renovation every ten years or so will have to be devised, as is said to be the rule now in the Western United States.

## BIBLIOGRAPHY

Brownlie, T. A. M.—Convoluted tube-wells for irrigation, *Agr. Jour. of India*, viii, 1913, p. 145.

Clarke, G.—Notes on improved methods of cane cultivation, Allahabad, 1919.

Gurtu, S. K.—Agriculture and irrigation: a problem of Economic development, *Agr. Jour. of India*, xiv, 1919, p. 541.

Howard, A.—The improvement of crop-production in India, *Jour. of the Roy. Soc. of Arts*, lxviii, 1920, p. 555.

Howard, A., and Howard, G. L. C.—The irrigation of alluvial soils, *Agr. Jour. of India*, xii, 1917, p. 185.

————————The saving of irrigation water in wheat growing, *Bull. 118, Agr. Research Institute, Pusa*, 1921.

Inglis, C. C.—Land drainage in irrigated tracts of the Bombay Deccan, *Agr. Jour. of India*, xiv, 1919, p. 781.

Irrigation in India: Review for 1921–2, Delhi, 1923.

Knight, J. B.—Work on the Manjri Farm for the years 1915–18, *Bull. 90, Dept. of Agr., Bombay*, 1918.

Main, T. F.—Irrigation in Sind, *Agr. Jour. of India*, xiii, 1918, p. 653.

Report of the Indian Irrigation Commission, Calcutta, 1903.

Report of the Indian Sugar Committee, Simla, 1921.

Roberts, W.—Some aspects of large estate farming in the Punjab, *Agr. Jour. of India*, xix, 1924, p. 141.

Triennial Review of Irrigation in India, 1918–21, Calcutta, 1922.

Vick, F. H.—The tube-well and its irrigation possibilities, *Agr. Jour. of India*, xi, 1916, p. 296.

## IV

# THE NITROGEN PROBLEM

After the regulation of the water-supply, the solution of the nitrogen problem is the next step in the development of Indian agriculture. While experience shows that crop-production can be materially developed by improving the water-supply, nevertheless the highest duty of water is only reached when we increase at the same time the supply of combined nitrogen. Both these must be adequate if the optimum results are desired. After arranging for water, the problem is gradually to increase the nitrogen supply by methods within the means of the cultivator.

As far as combined nitrogen is concerned, a rough condition of equilibrium has been reached in Indian agriculture. The annual gains and losses are about equal. This is indicated by the records of crop-production in the United Provinces from the time of Akbar down to the present day. For the last 300 years the producing power of the soil has not appreciably changed. If, therefore, it is desired to increase the yield per acre, either the losses of nitrogen must be reduced or the amount added to the soil must be increased or rendered more effective.

The losses of nitrogen in Indian agriculture are many and obvious. In the absence of a sufficient supply of firewood over a large part of the country, most of the cow-dung produced is burnt as fuel and, in the process, the nitrogen contained in it is lost to agriculture. Every year many tons of combined nitrogen are exported, chiefly in the form of oil-seeds, various food and other grains, tea, and animal products like hides and bones. The destruction of nitrates in the soil by incipient water-logging during the monsoon is another important source of loss. Practically none of this loss is made up for by the importation of nitrogenous fertilizers. Indeed most of the sulphate of ammonia from the coal-fields is not used in India, but is exported to Java and the Straits Settlements. Crude saltpetre, although manufactured in India, is not applied to the land by the cultivators to any great extent.

The first step in the consideration of the problem is to ascertain whether any of the losses can be avoided or reduced. We cannot advise the cultivator to give up burning cow-dung for fuel without providing some substitute. So far, no one has succeeded in accomplishing this. The export of seeds and other products containing nitrogen must obviously go on, otherwise the cultivator will soon find himself without money with which to pay his dues and purchase the necessaries of life. Incidentally, any large check in exports would soon set up an adverse trade balance, and a fall in the exchange value of the rupee would follow. The only source of loss which can be reduced is that brought about by drainage and local water-logging during the rains. This could be largely avoided by improving the grading of the fields and by providing a suitable system of surface-drainage. This loss is considerable, and its reduction is the first step in improving the nitrogen supply and thereby increasing crop-production.

The second step in the nitrogen question is how best to increase the amount and efficiency of the manures at the disposal of the cultivator. Several indigenous sources of combined nitrogen exist in India, namely nitrogen fixation, green-manures, oil-cakes, and organic residues. These must now be considered from the standpoint of rural India. One of the most important agencies in nitrogen fixation is the leguminous crop, whose root-nodule organisms work up the free nitrogen gas of the air into complex substances which the plant can use and which are partly left behind for succeeding crops. The cultivation of leguminous crops, therefore, means an important gain in combined nitrogen. Although it was not till 1888, after a protracted controversy lasting thirty years, that Western science finally accepted as proved the important part played by these crops in enriching the soil, nevertheless centuries of experience had already taught the cultivators of the Orient the same lesson. The leguminous crop in the rotation is everywhere one of their old fixed practices. There is one exception. When the engineers, by means of perennial canals, enabled the surplus population of the congested areas of the Eastern Punjab to conquer the desert, a system of exploitation of the virgin soil took place not unlike that in North America when the great Western movement occurred. On the new Canal Colonies there was no pressure of population, the soil responded to irrigation, and the old tradition of the need of the leguminous rotation to keep up the fertility of the soil was forgotten. The cultivation of wheat, cotton, and oil-seeds meant money. For a time nothing else seemed to matter. The time, however, is coming in the Punjab, as it came in North America, when this exploitation will have to stop, and when more attention will have to be paid to leguminous crops like gram and to the various fodder plants for preventing a further fall in production. The best method of employing the leguminous crop in increasing the fertility will then become a practical question. This was one of the subjects discussed during the symposium on the nitrogen problem in Indian agriculture held at the meeting of the Indian Science Congress at Lucknow in 1923. It is dealt with in detail in a subsequent chapter on the pulse crops of India (pp. 162–63).

There are other soil organisms besides those found in root-nodules which fix atmospheric nitrogen and thereby add every year large quantities of combined nitrogen to the soil. These derive their energy from the organic matter in the soil, and need a constant supply of oxygen for their activities. They also need time. Although the Indian cultivator knows nothing of *Azotobacter*—one of the organisms concerned in this nitrogen fixation—he is aware of the fact that a well-managed monsoon fallow, in which the surface is kept stirred, means a good wheat crop, and that the land when properly rested and tilled will manure itself. In Sind and Baluchistan, where the fallows extend over three years or longer and the fixation process is allowed more time, the results are still more striking. In Sind heavy crops of millets, which require large quantities of nitrogen, succeed one another without any manure beyond the intervening period of rest. Long fallows are all very well in sparsely populated tracts like Sind, but where the pressure of population is considerable their duration must be shortened. Can we speed up the fixation process and achieve, between two cold-season crops, a result comparable with a Sind fallow which takes at least two years and often more? Investigations are needed to answer this question, but there is some evidence for the view that, as the texture of the soil is improved by the addition of decayed organic matter, the amount of natural fixation increases considerably. If this idea is borne out by further experiments and experience, an important step in the solution of the nitrogen problem will have been made. It would then be possible to make the soil of India manufacture a great deal of the combined nitrogen it needs.

Besides the value of leguminous crops in the rotation, some of these can be used as green-manure. A very large amount of valuable work has been done on this subject in India in recent years, both on the time and manner of application and on the most suitable plant to use. In Madras and other parts of the country the outturn of rice has been markedly increased by this means. In the Central Provinces, the benefit of the sunn-hemp crop (*Crotalaria juncea* L.) used for this purpose has been cleverly divided between the wheat and rice areas. In the plains green-manure has been successful on light, open soils, but has often failed on heavier land except where small dressings of superphosphate have been added at the same time. Whether this result is due to an alteration in the physical texture of the soil brought about by the superphosphate or to the addition of a small amount of food material is an interesting question which so far has not been answered. Crucial experiments are needed to determine the precise effect of this manure. If the result is due to phosphorus, other manures which contain this element should produce the same result. In spite, however, of the large amount of work which has been done on green-manuring in India, there is a general feeling that much more remains to be accomplished and that little more than the fringe of the subject has been touched.

Another important indigenous source of combined nitrogen is oil-cake. The use of this material is spreading in India for money crops like sugar-cane and tobacco, and its value is well understood by the people. The amount available, however, is limited, as most of the oil-seeds grown are exported and the extraction of the oil takes place in Europe. The oil-cake does not return to India. Western industries need oil; Western agriculture needs oil-cakes both for stock and as manure.

To stop the loss of combined nitrogen involved in the export of the seed, it has frequently been suggested that the oil-pressing should be done in India, when the oil-cake would remain to improve the soil. Simple as this matter at first sight seems, a solution has not yet been found. It is much easier and cheaper to transport seed in gunny bags than to move oil without loss, deterioration, or contamination in drums or in tanks.[1]

It is in the better utilization of organic residues, both plant and animal, that an immediate improvement in the nitrogen supply seems possible. Little or nothing can be done at present in preventing the loss of nitrogen entailed in the burning of cow-dung. Sufficient use, however, does not appear to be made of the ashes, crop residues, leaves, and other forms of organic matter which are now applied to the land as manure. These are allowed to remain in heaps and are then carried to the fields. Simple methods, within the means of the cultivator, of composting these materials with earth and cow-dung on the lines adopted so successfully in China and Japan must be devised. In these countries intensive agriculture is practised, and the soil supports a much larger population than India without any importation of artificial manures. The greatest attention is paid in the Far East to the production of organic matter in the right stage of decomposition before it is applied to the fields. It is never used fresh and undecayed, as is often the case in India. Plant residues, in the hands of the peasantry of China, appear to be much more efficacious than they are in India, and it seems well worth while exploring the possibility of adapting many of the practices of the Far East to Indian agriculture. These have been graphically described by King in *Farmers of Forty Centuries*, and are readily accessible to the student of Indian agriculture.[2]

Two views are held in India as to the best way of increasing the nitrogen supply. One side holds that all that is necessary is to check the avoidable losses, to stimulate nitrogen fixation, to improve the process of green-manuring, to get the most out

[1] These questions are discussed in greater detail in the chapter on oil-seeds (pp. 147–50).

[2] The use of human excreta in preparing composts, as practised in China, is impossible in India.

of oil-cakes, and to introduce methods of composting organic matter similar to those in use in China and Japan. The other view is that such methods will not suffice, and that artificial manures are necessary in stimulating crop-production in India as they are in Western agriculture. There is an excellent discussion of this question in the recently published *Report of the Sugar Committee* in so far as it relates to the intensive cultivation of crops like cane. The Committee concluded that the most promising sources of combined nitrogen are: (1) the increased production of oil-cakes; (2) the development of the manufacture of sulphate of ammonia on the Indian coal-fields; and (3) the establishment of synthetic processes for obtaining combined nitrogen from the atmosphere. The case for the increased use of artificial nitrogenous fertilizers has been put with great clearness and force, and deserves the fullest consideration. No particular difficulties are likely to be encountered in the manufacture in India of synthetic nitrogen products like calcium cyanamide and ammonium sulphate by means of the Haber process. During the war a great deal of the nitrogen in the tons of explosives used was obtained from the atmosphere by these processes, so that the necessary experience and plant can easily be imported. Coal, lime, and water-power are available in India, so that, if nitrogenous fertilizers of this class are found to be necessary, they can be made in the country. Two difficulties, however, will have to be got over before they can be taken up widely by the cultivators. Adulteration in the bazaars must be prevented, and a Fertilizers Act will be desirable for the protection of the peasantry. The price charged must be low, as substances like cyanamide and sulphate of ammonia produce their greatest effect on the crop to which they are applied, and do not influence the texture of the soil by any residue of organic matter, as is the case when materials like oil-cakes and green-manures are used.

Before any decision can be arrived at as to the best method of solving the nitrogen problem, a good deal of careful experiment will be necessary. All the somewhat fragmentary work which has been done on this question should be brought together, at two or three of the existing experimental farms, and should be woven into definite projects. Their aim should be to ascertain the cheapest and easiest transition from the present extensive methods of agriculture to an intensive system, designed to give the highest possible yields. Typical areas of land of suitable size should be taken up and worked as experiments in rural economy according to plan. Careful records of cost and of production, as well as a nitrogen balance-sheet, are of course indispensable. An attempt should also be made to ascertain whether, as seems possible, the amount of free nitrogen fixed by the soil organisms increases as the land becomes richer in organic matter. At least two methods of manuring should be tried at these centres. One might usefully follow the lead of China and Japan, and make the fullest use of surface-drainage, nitrogen fixation, green-manures, and the various forms of organic residues. Another might proceed on the lines suggested by the Sugar Committee, and employ sulphate of ammonia from the coal-fields and synthetic nitrogen compounds like calcium cyanamide. It would be exceedingly interesting to compare the cost, results, and the nitrogen balance-sheets in the two cases. From these it ought to be possible to formulate the leading principles which should be followed in the practical solution of this matter. At two centres in India—Shahjahanpur in the plains and Manjri in the Deccan—a beginning has been made in the direction indicated, and a great deal of useful work has been done in working out intensive methods of sugar-cane cultivation, based on the proper use of organic matter, water, and improved cultivation. The ideal experiment station, devoted solely to the practical solution of the nitrogen problem, remains, however, to be devised.

## BIBLIOGRAPHY

Burt, B. C.—Some experiments in manuring for double cropping, *Agr. Jour. of India*, xii, 1917, p. 536.

Clarke, G., Banerjee, S. C., Naib Husain, M., and Qayum, A.—Nitrate fluctuation in the Gangetic alluvium and some aspects of the nitrogen problem in India, *Agr. Jour. of India*, xvii, 1922, p. 463.

Clouston, D.—Urine earth as a manure, *Agr. Jour. of India*, v, 1910, p. 262.

————Manures in their relation to soils and crop-production in the Central Provinces, *Agr. Jour. of India*, xiv, 1919, p. 101.

Couchman, M. E.—Leaf-manure in South Canara, *Agr. Jour. of India*, iii, 1908, p. 232.

Dobbs, A. C.—Green-manuring in India, *Bull. 56, Agr. Research Institute, Pusa*, 1916.

Howard, A.—Soil-erosion and surface-drainage, *Bull. 53, Agr. Research Institute, Pusa*, 1916.

Howard, A., and Howard, G. L. C.—Drainage and crop-production in India, *Agr. Jour. of India*, xiv, 1919, p. 377.

Hutchinson, C. M.—Nitrogenous fertilizers: their use in India, *Agr. Jour. of India*, xiv, 1919, p. 203.

————————Nitrogen fixation in Indian soils, *Agr. Jour. of India*, xiv, 1919, p. 215.

King, F. H.—Farmers of forty centuries, or permanent agriculture in China, Korea, and Japan, Madison, Wisconsin, 1911.

Knight, J. B.—Work on the Manjri Farm for the years 1915–18, *Bull. 90, Dept. of Agr., Bombay*, 1918.

Lander, P. E., Wilsdon, B. H., and Mukand Lal, M.—A study of the factors operative in the value of green-manure, *Bull. 149, Agr. Research Institute, Pusa*, 1923.

Main, T. F.—Irrigation in Sind, *Agr. Jour. of India*, xiii, 1918, p. 653.

Report of the Indian Sugar Committee, Simla, 1921.

The Nitrogen Problem in Indian Agriculture, *Proc. Tenth Indian Science Congress*, Asiatic Soc. of Bengal, Calcutta, 1924, pp. 243–61.

Wilsdon, B. H., and Barkat Ali.—Nitrogen fixation in arid climates, *Soil Science*, xiv, 1922, and *Agr. Jour. of India*, xviii, 1923, p. 158.

# V

# THE ALKALI PROBLEM

ONE of the most striking features of Northern India is the vast expanse of barren (*usar*) lands which alternate with the fertile and populous areas. Such alkali lands are common in Oudh, in the Province of Agra, in various parts of the Punjab and Sind, and on the Western Frontier. They also occur on the Nira Canal in the Deccan and in the District of Kaira in Bombay. Generally speaking, however, alkali interferes with crop-production in India to an important extent only on the alluvium of the Indo-Gangetic plain, particularly in North-West India. The existence of these infertile areas involves a serious loss of land revenue. Their reclamation would help materially in solving the problem of over-population in the tracts in which they occur.

The formation of alkali land in India is intimately connected with irrigation. Except on very permeable soils, any tendency towards over-irrigation is almost certain to increase the amount of alkali salts. Where these salts are present in injurious amounts, they appear on the surface in the form of snow-white or brownish-black incrustations, known as *reh* or *kallar*. The former (white alkali) consists largely of the sulphate and chloride of sodium, the latter (the dreaded black alkali) contains sodium carbonate in addition, and owes its dark colour to the fact that this salt is able to dissolve the organic matter of the soil. According to Hilgard, sodium carbonate is formed from the sulphate and chloride in the presence of carbon dioxide and water. The action is reversed in the presence of oxygen.

The occurrence of minute quantities of alkali salts in the soil has no injurious effects on crops or on the soil organisms. It is only when the proportion increases beyond a certain limit that they first interfere with growth and finally prevent it altogether. Leguminous crops are particularly sensitive to alkali, especially where this contains carbonate of soda. The action is a physical one, and depends on what is known as the osmotic pressure of solutions, which increases with the amount of dissolved substance. For water to pass readily from the soil into the roots of plants, the osmotic pressure of the cells of the root must be considerably greater than that of the soil solution. If the soil solution becomes stronger than that of the cells, water will pass backwards from the roots to the soil and the crops will wither. This state of affairs actually occurs when the soil becomes charged with alkali salts beyond a certain point. The crops are then unable to take up water, and death results. The roots behave similarly to a plump strawberry when placed in a strong solution of sugar. Like the strawberry, they shrink in size because they have lost water to the stronger solution outside. Too much salt in the soil, therefore, tends to make irrigation water useless.

The production of these salts in injurious amounts is found to be closely associated with the texture of the soil. If the soils are open, permeable, and well aerated, alkali salts are absent. On the other hand, deep layers of stiff, heavy, poorly-aerated clays are almost certain to be affected by alkali when such areas are brought under perennial irrigation. Alkali also appears in the stiffer soils when accumulations of stagnant water raise the subsoil water-level and when the surface-drainage is held up by roads or by embankments of various kinds. Recent investigations in Mesopotamia have furnished interesting confirmation of the connexion between alkali and poor soil-aeration. Intensive cultivation is only met with in that country when the soils are permeable and when the natural drainage is good. Where the drainage and aeration are poor, the alkali question becomes acute.

The solution of the alkali problem concerns a great deal more than the reclamation of the existing barren areas, important as this is. In the soils of the alluvium there appears to be a great deal of transient alkali which comes and goes according to the season and to the method of soil management adopted. This frequently diminishes the yield. On soils like those at Pusa and at Quetta the land appears to hover on the verge of alkali formation. The greatest care is necessary to maintain its texture and permeability so as to keep this tendency in check.

In cases where the subsoils are stiff this is particularly necessary. Unless deep-rooting crops like sanai (*Crotalaria juncea* L.), rahar (*Cajanus indicus* Spreng.), and lucerne are frequently included in the rotation, semi-impermeable subsoil pans result which lower the fertility. The formation of this pan appears to usher in the alkali phase.

The importance of the alkali problem in India has been recognized for many years, and much time and money have been devoted to its solution. From the time of the *Reh* Committee of 1877 down to the publication of Moreland's summary of this question in the *Agricultural Ledger* of 1901, the literature contains numerous references to the investigations in progress on *reh* and *kallar*. During the present century the work has been continued, mainly in connexion with the salt lands of the Nira valley in Bombay and with the reclamation experiments at Daulatpur in Sind and at Narwala near Lyallpur. These various investigations have led to the accumulation of a large volume of data which has placed the nature and composition of *reh* beyond dispute. The impermeability of many of the alkali soils, and particularly of those containing sodium carbonate, has also been demonstrated, as well as the frequent association of beds of concretionary limestone (*kankar*) underneath *usar* soils. In some cases definite information has been recorded which proves that alkali lands have arisen as a direct result of canal irrigation, and that the evil is increasing every year. In the Nira valley, for example, alkali was unknown before the canal was constructed, but now covers thousands of acres and is steadily increasing. The greatest amount of attention has been paid to the reclamation of these barren soils, and much energy has been bestowed on the discovery of a practical solution of the problem. The results, however, have been for the most part negative. Where bulky organic manures have been available in large quantities as a by-product, as in the case of the dairy near Aligarh, permanent benefit has been obtained by their use, and alkali lands have in this way been brought into cultivation. A certain amount of success has attended the use of gypsum (sulphate of lime), which transforms black alkali into the less harmful sulphate of soda. In exceptional cases, mixing alkali soils with sand is successfully practised by the cultivators, while on the Western Frontier the growth of lucerne is recognized as the best means of dealing with soils slightly affected by white alkali. This deep-rooting fodder crop improves the texture and the aeration of the soil, and brings about a natural cure which persists until the permeability is again ruined by frequent irrigation. At Narwala in the Chenab Colony, alkali soils have been temporarily reclaimed by mole-drainage, combined with washing out the salts with large quantities of irrigation water. Positive results have also been obtained on a small scale in the Nira valley by the use of primitive drains, but the cost is likely to be prohibitive. The universal reclamation method in use in the United States, namely subsoil drainage, has not succeeded on the alluvium, as the drains either silted up or did not function on account of the impermeable character of many of the *usar* soils.

A critical examination of these investigations reveals one very general weakness, namely a want of attention to the conditions which are necessary for the formation and accumulation of alkali salts. Almost all workers on this problem appear to have been obsessed by the importance of reclamation and by the need of obtaining a practical result. It is generally supposed that alkali soils are the natural consequences of a light rainfall, insufficient to wash out of the land the salts which always form in it by progressive weathering of the rock powder of which all soils largely consist. Hence alkali lands are considered to be a natural feature of arid tracts, such as parts of the Punjab and Sind where the rainfall is very small. Such ideas on the origin and occurrence of alkali lands in India, however, do not correspond with all the facts. The rainfall of the United Provinces of Agra and Oudh, although often deficient, is nevertheless considerable, particularly in Oudh, and is certainly adequate for the removal of the comparatively small quantities of soluble salts found in the *usar* plains. In North Bihar the average rainfall, in the sub-montane tracts where alkali patches are common, is from fifty to sixty inches. Arid conditions, therefore, are not essential for the production of these alkali salts; heavy rainfall does not always remove them. What does appear to be a necessary condition is defective soil-aeration. Wherever the air-supply is cut off by the constant surface irrigation of stiff soils with a tendency to impermeability, by the accumulation of stagnant subsoil water, or by interference with the surface-drainage, alkali salts sooner or later appear. Almost any agency which interferes with soil-aeration for a long period will produce alkali land. In North Bihar old roads and the sites of bamboo clumps and certain trees, like the *pipul* (*Ficus religiosa* L.) and the tamarind (*Tamarindus indica* L.), produce alkali some time *after* they are brought into cultivation. In such cases the alkali condition takes months to develop. If these areas are examined, they are frequently found to contain the bluish-green markings which are associated with the activities of those soil organisms which are able to live in badly drained soils without a supply of free oxygen. These anaerobic organisms appear to bring about a phase of oxygen hunger (reduction) in the soil which involves the formation of substances like sulphuretted hydrogen and the metallic sulphides. When circumstances alter and these soils come in contact with air and water, oxidation takes place with the ultimate formation of the salts of alkali land—the sulphate, chloride, and carbonate of sodium. Thus the first step in the formation of alkali appears to be intense reduction during which most of the combined oxygen in the soil is consumed. It is the subsequent oxidation of the products of this reduction which forms and maintains the alkali salts. When the reductive phase has been long continued and the oxidation is only partial, the alkali forms slowly and accumulates in the soil. When copious aeration and drainage can be provided, the oxidation is completed and the salts are removed. After this there is no further production of alkali unless aeration is again interrupted. The possibility of the existence of these reducing conditions and their importance in the formation of alkali lands has been pointed out by Cameron and others in the United States and has recently received interesting confirmation in India. In the alkali zone in North Bihar wells have to be left open to the air, otherwise the water is contaminated by sulphuretted hydrogen, thereby indicating a well-marked reductive phase in the deeper soil layers. In a subsoil drainage experiment in the Nira valley,

Mann and Tamhane found that the salt water which ran out of these drains soon smelt strongly of sulphuretted hydrogen, and a white deposit of sulphur was formed at the mouth of each drain, proving how strong were the reducing actions in this soil. Here the reduction phase in alkali formation was actually demonstrated. After drainage and aeration were established, the conditions necessary for alkali production were removed and the original texture and fertility of the soil were restored. Somewhat similar results have been obtained at Quetta in the case of the dense loess soils met with there. Here again the first stage in alkali formation appears to be a reductive phase. The bluish-green and brownish markings, characteristic of badly-drained soils, are met with a few inches below the alkali patches. At the surface there is a white incrustation. If such soils are copiously aerated and the alkali salts are at the same time washed out, they become very fertile and remain so. These indications, imperfect as they are, nevertheless point to the supreme importance of further investigation of the origin of alkali lands in India and the discovery of the conditions which are necessary for their formation. Once this is accomplished, the questions of prevention and cure might enter on a new phase.

When the origin of alkali is understood, a great step forward will have been made and another principle will have been added to agricultural science on alluvial soils. It will then be possible to undertake a scientific examination of the bases of perennial irrigation, as practised at the present day in the plains, and to prevent any misuse of the method. Such an investigation is particularly necessary at the present time in view of the pending developments in canal irrigation in Sind and in Oudh, where the texture of the soils is distinctly heavier and the permeability less than in the areas hitherto commanded by perennial canals. If, as appears to be the case, poor soil-aeration is found to be an important factor in the production of alkali, we must expect that the institution of perennial irrigation on the stiff soils on the left bank of the Indus and in parts of Oudh will lead to the formation of alkali land on a large scale in a comparatively short time, unless means are taken to maintain soil-aeration. It must always be remembered that the ancient irrigators never developed any efficient method of perennial irrigation, but were content with the basin system, a device by which irrigation and drainage can be combined. In his *Irrigation and Drainage* King concludes an interesting discussion of this question in the following words, which deserve the fullest consideration on the part of the irrigation authorities in India : ' It is a noteworthy fact that the excessive development of alkalies in India, as well as in Egypt and California, are the results of irrigation practices modern in their origin and modes, and instituted by people lacking in the traditions of the ancient irrigators who had worked these same lands thousands of years before. The alkali lands of to-day, in their intense form, are of modern origin, due to practices which are evidently inadmissible, and which, in all probability, were known to be so by the people whom our modern civilization has supplanted.'

## BIBLIOGRAPHY

Barnes, J. H., and Barkat Ali.—Alkali soils, some bio-chemical factors in their reclamation, *Agr. Jour. of India*, xii, 1917, p. 368.

Dorsey, C. W.—Alkali soils of the United States, a review of literature and summary of present information, *Bull. 35, Bureau of Soils, U.S. Department of Agriculture*, 1906.

Harris, S. F.—Soil Alkali, its origin, nature, and treatment, New York, 1920.

Henderson, G. S.—Notes on practical salt land reclamation, *Bull 91, Agr. Research Institute, Pusa*, 1920.

Hilgard, E. W.—Soils, New York, 1906.

Howard, A., and Howard, G. L. C.—The agricultural development of Baluchistan, *Bull. 119, Agr. Research Institute, Pusa*, 1921.

King, F. H.—Irrigation and drainage, London, 1900.

Leather, J. W.—Reclamation of *reh* or *usar* land, *Agricultural Ledger*, Nos. 7 and 13, 1897.

——————Investigations on *usar* land in the United Provinces, Allahabad, 1914.

Mann, H. H., and Tamhane, V. A.—The salt lands of the Nira valley, *Bull. 39, Dept. of Agr., Bombay*, 1910.

Moreland, W. H.—Reh. An account of the attempts which have been made to utilize the upland barren lands (*usar*) of the United Provinces for profitable purposes, *Agricultural Ledger*, No. 13, 1901.

Sen, J. N.—Some observations on the occurrence of infertility under trees, *Agr. Jour. of India*, xii, 1917, p. 390.

Tamhane, V. A.—Investigations into the nature of the salt lands in Sind, *Bull. 96, Dept. of Agr., Bombay*, 1921.

Watt, G.—Dictionary of Economic Products of India, vi, pt. 1, 1892, p. 400.

——————The Commercial Products of India, London, 1908, p. 51.

Webster, J. F.—Alkali lands in Iraq, *Memoirs 1 and 5, Dept. of Agr., Mesopotamia*, 1921 and 1922.

# PART II

# THE CROP

INDIA is a land of small-holders devoted to the raising of crops. These are of two kinds. First in area and importance are the food crops—the cereals and pulses which feed the population. The second group comprises the money crops, by which the cultivator pays the land revenue and purchases the necessaries of life. The surplus produce, which remains after the needs of the country are satisfied, is exported.

The field of work in the improvement of crops in India is immense. More efficient varieties are needed to replace those now cultivated. More intensive methods of agriculture, within the means of the people, have to be devised. It will be seen in the following chapters that progress has been made in both these directions. Such advances, however, must of necessity be slow, as they have to proceed from the basis of small holdings cultivated by a peasantry for the most part in debt.

The introduction of improved varieties of crops and the provision of seed at moderate rates is the first step in the uplift of rural India. In the replacement of the country crop by the new kinds, the cultivator is not put to any great expense; the increase in production improves his position and also serves to establish confidence. The next stage is the gradual improvement of the soil conditions so that the full benefit of the new varieties can be obtained.

Two conditions must be satisfied in the work of crop-improvement. The first is an adequate realization of the working conditions of the average cultivator, so that all proposals for his benefit may be practicable. The second is the need of a thorough study, at the Experiment Stations, of the problems to be attacked. Anything brought to the notice of the people should be the best that can be devised in the present stage of knowledge. A constant succession of new varieties or frequent modifications of some process already introduced only bewilder the cultivator and tend to lower the confidence of the public in the work.

In the following fourteen chapters the chief crops of India have been discussed and an attempt has been made to describe the progress already made and to indicate some of the directions in which further work is desirable. A few general questions, such as the significance of root-development, methods of seed-distribution, and the need for the further study of disease, have also been considered.

# VI

# THE ECONOMIC SIGNIFICANCE OF ROOT-DEVELOPMENT

THE most important agent in crop-production is the plant itself. In all countries agriculture ultimately rests on crops; everywhere the plant is the real centre of the subject. Improvements of the soil are in a sense subsidiary, and are undertaken with the sole object of increasing the activities of crops and of getting more work done by the plant. The soil serves to anchor the crop and to provide water and certain food-materials for the green leaves, which the plant absorbs by means of an extensive root-system. The roots are the means by which soil and plant are brought into gear. The more the nature and extent of this gearing is examined, the more important and significant it becomes.

Notwithstanding the enormous number of field experiments and variety trials which have been carried out in the past in all parts of the world, comparatively little is known of the root-systems of agricultural crops. Investigators often forget that they see only half the crop under experiment and that the most important portion, the root-system, is below ground. A change, however, has recently taken place, and much more attention is now being paid to root-development. A beginning was made some years ago in India, and the subject is being energetically explored in the United States by the workers associated with the Carnegie Institution of Washington, D.C. During the past year Weaver and his colleagues have published an important monograph entitled *Development and activities of roots of crop plants* which marks a definite stage in progress. In this paper, interesting studies on the root-development of cereals, potatoes, and fodder crops are described which bring out clearly the great depth to which the roots of crops penetrate. In practically all cases where these were exposed, the total development below the top layer of cultivated soil was as great and usually much greater than that in the surface soil itself. For example, at ten different stations in the prairie areas of Nebraska, Kansas, and South Dakota, the bulk of wheat roots penetrated to three feet eight inches, while most of the roots of rye reached four feet. These facts contradict the current statements in the literature on soils to the effect that only the top six or eight inches of soil is suited to plant life and that the subsoil plays only an indirect part in plant nutrition. Weaver found that the deeper layers of soil are not only suited to plant life, but play an exceedingly important part in root-development. The significance of these facts will be evident when it is remembered that only the upper five or six inches of soil are directly influenced by cultivation and manure, and that the soil investigations of the past have been concerned mainly with samples taken to a depth of nine inches. The first condition in future soil studies will be to ascertain the facts relating to the root range of crops, and then to look at the soil from the plant's point of view.

The wide range in the soil and agricultural conditions of India render this country particularly suitable both for the investigation of the root-system of crops and also for working out the agricultural significance of the differences encountered. Between the black soils of the Peninsula and the alluvium of the plains various intermediate soil types occur. Large deltaic areas of rice land are found round the shores of the Bay of Bengal, while in the Assam valley acid soils occur, the permeability of which is liable to deterioration. In the Central Provinces and South India laterite soils abound characterized by a high degree of porosity and correspondingly good aeration. Further, the agriculture in many of these areas is ancient, there have been few innovations, and the soil conditions have had time to impress themselves on the varieties of crops cultivated. A condition of equilibrium between the type of plant and the soil has been obtained, as there has been ample time for the operation of natural selection. When we compare the root-system of linseed from the black soil areas with that of the varieties grown on the Gangetic alluvium, striking differences appear. The roots produced on the black soils are deep, somewhat sparse, and are well adapted to ripen the plant quickly with the minimum of moisture. The type of gearing fits the soil. On the alluvium, where moisture is more abundant and where the aeration of the subsoil is poor, the root-system is superficial but at the same time well developed. On the intermediate types of soil, which occur near the line of the Jumna and the Ganges, linseed produces a type of root about half-way between that of the black soils and of the alluvium. Once more the root-system is found to fit the soil type. Further, when we grow side by side on the alluvium these three classes of linseed, there is little or no adaptation of the roots to the new conditions, but the three types behave very much as they would in their native habitat. The deep, sparse root-system of the black soil areas is developed in the alluvium, although it is fatal to the well-being of the crop. When the experiment is reversed and the types which suit the alluvium are grown on the black soils, there is again little or no adaptation to fit the new conditions. The linseed crop consists of a large number of varieties which differ from one another in all sorts of characters, including the extent and distribution of the roots. The root-systems of the varieties are just as characteristic and just as fixed as the differences in the seed and other above-ground characters of these plants. A similar state of affairs obtains in other crops, like wheat, *khesari* (*Lathyrus sativus* L.), *patwa* (*Hibiscus cannabinus* L.), and the opium poppy, and is probably universal all over India. These differences between the root-systems of varieties have important consequences. We have seen that the soil and the crop are brought into gear by means of the roots of the plant. If we aim at obtaining the highest possible yield, it follows that this gearing must be efficient. We cannot remove the soil of a locality and substitute one of better quality from elsewhere. One element in the system is therefore fixed. As regards the other unit—the plant—nature has provided us with a considerable amount of scope. We can select among the various types of root-system available and find the one which will connect the plant and the soil in the most efficient manner. In this way only can the crop make the best use of the soil. A great deal of the work in connexion with the improvement of varieties consists in changing gear.

The root-system has other uses in agriculture besides connecting the crop to the soil. It is an important agent in cultivation and in maintaining fertility. Recent investigations in India indicate that the best way of maintaining the physical condition of the subsoil is by means of root-systems carefully selected for the purpose. It is well known that one of the frequent consequences of the constant cultivation of the upper soil is the formation of more or less impermeable pans just below the plough-sole. These interfere with the permeability and aeration of the subsoil and lower the yield. In Western agriculture these are often broken up by mechanical means with excellent results. The Indian cultivator, however, cannot command expensive power for such deep cultivation, and therefore cannot make use of the brute force methods of the Occident. He has, however, unconsciously selected an excellent implement for the purpose, which not only does the work for nothing, but at the same time yields a dividend. The subsoiler of the cultivators of the alluvium is a plant—the crop known as *rahar* (*Cajanus indicus* Spreng.)—which is grown in the rotation about once in four years. This plant produces a very deep and extensive root-system, the branches of which are much thicker and more robust than those of the cereals. They penetrate and break up ordinary pans with ease. The decay of the roots leaves numerous passages in the deeper layers for water and air, and in this way permeability and aeration are improved. At the same time the subsoil is provided with organic matter for nitrification. No implement yet designed can accomplish the work of the *rahar* crop on the small holdings of the alluvium. The soils are silt-like, and therefore close in texture. The subsoil cultivation must be thorough, it must cost little or nothing, and, moreover, the level of the surface must not be disturbed. The proper maintenance of the condition of the subsoil of all alluvial soils by means of deep-rooting crops should never be forgotten when the desert areas of North-West India are being conquered by the canal. At first many of these soils possess great natural permeability, and heavy crops of wheat and cotton can be raised for some years. After a time, however, the condition of the subsoil deteriorates, root penetration decreases, and the crops need more and more irrigation water. The inclusion of deep-rooted crops in the rotation would help to avoid these troubles, and every effort should be made to induce the cultivators to grow them. In a sense they are almost as important as the canal itself.

The necessity for the cultivation of deep-rooting crops for keeping the subsoil in a proper condition is not confined to the plains. It is often an urgent matter in the case of permanent crops like tea and rubber. Here the forest canopy, which is made up of a large number of species, differing widely in their root-development, has to be removed so that one kind of plant with a uniform type of root can be grown instead. As in the conquest of the desert by canal irrigation, everything goes well at first till the supply of organic matter in the upper soil is used up and the permeability of the subsoil is lost. Trouble then begins. The yield falls off, and various pests begin to make their appearance. As the land is already planted, it is not possible to do much by methods of cultivation. Something can be accomplished by manuring the surface soil, and in some cases the pests can be kept in check by spraying and other means. At the best these methods are little more than palliatives, and often do not go to the root of the matter. It is probable that these troubles are due, in part at least, to the fact that the subsoil is never penetrated by roots and that the various consolidating agencies are allowed free play. In many cases, when leguminous shade trees are grown among tea and cocoa, the plantation markedly benefits. It is more than probable that the good results are partly due to the subsoiling effect of the root-systems of these trees. The great differences of opinion which exist on the value of shade trees might be explained if we knew more about the root-development of these trees and its relation to the subsoil. It would not be surprising if the failure of certain trees is traced to the use of species whose root-systems are useless in breaking up subsoil pans and which only compete with the roots of the crop for water, nutrients, and air. The whole question of shade trees should be re-examined from this point of view. The first step is a thorough exploration and study of the root-systems of the various trees. The information gained would enable the most effective root-system to be selected for any particular soil type. The choice of tree would then be less a matter of chance or fancy than is now the case. It is not impossible that the many pests and obscure diseases which are now associated with the rubber tree in the Federated Malay States and in Ceylon may be mitigated after a close study of the subsoils and their amelioration by means of suitable shade trees. As is well known, there was nothing wrong with the vegetation of these areas when the planting was left to Nature. At the present time no single species seems to have a larger number of diseases than the rubber tree. Perhaps the use of suitable shade trees will reduce their number and importance in the future.

These examples by no means exhaust the profitable directions in which a study of the root-systems of agricultural crops can be prosecuted. Up to the present, rotations of crops, the application of manures and of irrigation water, the preparation of the land for the seed, as well as methods of inter-tillage, have all been settled in the most empirical fashion. Such knowledge as exists has either arisen from experience or has been obtained indirectly as the result of innumerable costly experiments. A shorter and more scientific road seems indicated by the study of the root-system and by following up its development on the various soil types and under different conditions. In this way the investigator will begin to see crop-production from the plant's point of view, and will be able to make use of the crop itself for indicating the direction of future improvement. At the same time, the various growth factors such as soil moisture, soil air, nutrients, and soil temperature will fall into their proper place in the system when the time comes for considering the results of experiments and for drawing conclusions.

## BIBLIOGRAPHY

Bergman, H. F.—The relation of aeration to the growth and activity of roots and its influence on the ecesis of plants in swamps, *Ann. Bot.*, xxxiv, 1920, p. 13.

Brenchley, W. E., and Jackson, V. G.—Root development in barley and wheat under different conditions of growth, *Ann. Bot.*, xxxv, 1921, p. 532.

Howard, A., and Howard, G. L. C.—The economic significance of the root development of agricultural crops, *Agr. Jour. of India* (*Indian Science Congress Number*), xii, 1917, p. 17.

Howard, A., and Howard, G. L. C.—Drainage and crop-production in India, *Agr. Jour. of India*, xiv, 1919, p. 377.

————————————Some aspects of the indigo industry in Bihar, *Mem. of the Dept. of Agr. in India (Botanical Series)*, xi, 1920, p. 1.

Mann, H. H., and Hutchinson, C. M.—Green-manuring in tea culture in India, *Bull. 2, Indian Tea Association*, 1906.

Miller, E. C.—Comparative study of the root systems and leaf areas of corn and the sorghums, *Jour. of Agr. Research*, vi, 1916, p. 311.

Shepperd, J. H.—Root systems of field crops, *Bull. 64, North Dakota Expt. Station*, 1905.

Ten Eyck, A. M.—A study of the root systems of cultivated plants grown as farm crops, *Bull. 43, North Dakota Expt. Station*, 1900.

Venkatraman, T. S.—Sugar-cane root systems—studies in development and anatomy, *Agr. Jour. of India*, xvii, 1922, p. 381.

Weaver, J. E.—The ecological relations of roots, *Publication No. 286, Carnegie Institution of Washington*, 1919.

Weaver, J. E., Jean, F. C., and Crist, J. W.—Development and activities of roots of crop plants ; a study in crop ecology, *Publication No. 316, Carnegie Institution of Washington*, 1922.

# VII

## THE IMPROVEMENT OF VARIETIES

WHEN we examine the agricultural products of India, it is at once evident that crops are of far greater importance than animals. This state of affairs is likely to continue, as the pressure of population will probably increase rather than diminish. The people are largely vegetarian, while seeds and fibres are the most important articles of export. The Indian cultivator is a grower of crops, his live stock are mere aids in cultivation and in the feeding of his family. The country does not export meat, wool, or dairy products.

Notwithstanding the paramount place of the crop in Indian agriculture, the average cultivator rarely devotes much attention to the seed he sows. He is too poor to pay the necessary premium for improved seed, consequently seed-merchants, in the European sense, do not exist. An Indian Garton has not yet established himself, for the simple reason that there is no profitable market for his productions. With a few exceptions, there is no attempt to keep the various varieties separate, so that mixed crops are the rule. Most of the mixtures grown are low yielders of no particular quality and with little power to respond successfully to improved soil conditions. This want of attention to the variety and the frequent absence of high quality in the final product all correspond with the general low standard of agriculture. It tends not only to a low level of production but also to a lack of uniformity in the produce. Want of uniformity made no difference as long as India was mainly concerned with the feeding of her population. This state of affairs no longer obtains. The country must now export raw materials and easily manufactured articles like gunny bags to pay for imports. More and more attention is also being paid to working up raw products like cotton in India itself. Modern industry, whether Indian or European, needs a uniform product. Not only must the consignments be up to sample, but the raw produce should not change materially from year to year. In this way a reputation is established, and the work of the shippers and manufacturers is facilitated. The mixture of varieties which occurs in the fields of the cultivators and the want of attention to the seed-supply run counter to this demand. Serious loss of uniformity in the produce is the result. Further admixture often takes place after the product has left the threshing-floor or the field. The consequence is that Indian produce lacks that uniformity and evenness which modern industries demand. It does not fall into a definite grade, and therefore does not command much confidence or a high price. The advantage of a grade is not all on the side of the merchant and the manufacturer. The growers benefit by the enhanced price which eventually results. That it pays all concerned to attend to such matters is proved by the care devoted in Canada to the grading of such a comparatively cheap product as wheat.

These matters have long been recognized, and the trade is taking steps to protect itself against adulteration. The export of raw jute is regulated by the system of marks and contracts evolved by the Calcutta Baled Jute Association. Indian wheat and other seeds are now shipped on a pure basis, and the adulteration with earth, which was the rule twenty years ago, is no longer permitted. At the suggestion of the Central Cotton Committee, a Cotton Transport Act, designed to prevent the fraudulent adulteration and improper description of cotton, has been passed and has been put into force in the Broach area. Further measures, dealing with the marking of cotton bales and the regulation of gins and presses, are now under consideration. These steps in the encouragement of legitimate trade in raw products will be greatly facilitated as the cultivator is induced to pay more attention to the variety.

Three lines of advance are possible in improving the variety so that more produce and a more even product can be obtained. These are acclimatization, selection, and hybridization.

The simplest method of improving a crop is to import better seed from other countries and to replace the existing varieties by a superior type. Much time has been devoted to this work in India during the last hundred years, but the results obtained have been disappointing. Many attempts have been made to establish American cotton in various parts of the country. It is true that the area under this variety has increased in the Punjab during recent years, but the quality of the lint appears to have deteriorated and the fibre is now deficient in strength. American cotton is also grown on a small scale in the United Provinces near Cawnpore and in the Dharwar cotton tract in Bombay. Egyptian cotton has been tried in Sind, but does not suit the local conditions. Attempts have been made to improve the wheat and tobacco crops by the introduction of exotics, but in no case have these been taken up enthusiastically by the cultivators over large tracts of country. One reason of this general want of success appears to be due to the severe conditions imposed by the Indian climate. Both in the monsoon and, to a less extent, in cold weather crops, the growth period is limited and there is no great latitude either at sowing time or at harvest. In India the crop must establish quickly and must ripen quickly. Exotic varieties often fail because of the slowness with which the seedlings establish compared with the rate at which the indigenous kinds pass through the early stages. This is well seen in the case of tobacco. American types, although suitable in other respects, linger over the seedling stage and so miss the short growth period and begin to ripen too late in the season. The same disadvantage is shown by American cotton, which only does well if sown sufficiently early to establish itself before the monsoon. Exotic varieties always seem to require favoured treatment, and when this cannot be provided they have not proved a success.

The application of modern methods of selection and hybridization to the mixed crops now grown has proved much more successful than the acclimatization of exotics. These methods are the natural consequences of the enormous complexity of the Linnean species. In the wider sense the species contains a number of distinct varieties, which again are not uniform, but consist of numerous individual units which breed true. These are the elementary species, unit species, or pure lines. The wheat crop of the world, for example, consists of many thousands of elementary species, all differing slightly from each other and breeding true. The first step in applying selection methods to a crop in India is a systematic botanical survey of the forms met with in the various tracts and their classification into species, varieties, and elementary species. This clears the ground and gives the investigator a supply of raw material with which to work. Matters are plain sailing up to this point. The next step is more difficult, as it consists of a comparative study of the elementary species and the successive elimination of inefficient forms. After a few years' work, only a small number of promising kinds remain. A great deal obviously depends on the judgement and insight shown by the investigator in making the final selection for trial under cultivators' conditions. When he is fortunate enough to discover a really satisfactory unit species and when an adequate seed-supply and suitable methods of seed-distribution have been organized, the replacement of the inferior country mixtures by a uniform type of higher yield and uniform quality is only a matter of a few years. Examples of such improvement have been carried out in several crops, of which cotton and wheat are typical examples. In the Central Provinces the botanical composition of the cotton crop showed that one of the constituents of the principal mixture grown by the people yielded more lint per acre and possessed a higher ginning percentage than any of the others. This is a white-flowered variety, known as *Roseum*, which was found to do well all over the Province. Suitable methods of seed-distribution, in which the people readily co-operated, were rapidly set in motion. In 1917 the area under *Roseum* was 700,000 acres, and the increased profit was at least Rs. 15 an acre. In the case of wheat a botanical examination of the Indian crop, commenced in 1905, has led to the selection of a number of improved wheats, of which three—Pusa 12, Pusa 4, and Punjab 11—now cover over a million acres. The increased profit to the growers is at least Rs. 15 an acre. The annual dividend on these two pieces of work is already very great and is rapidly increasing.

After the possibilities of improvement by selection have been exhausted, further progress can be made by means of hybridization. By this process, unit species are crossed for the purpose of bringing about new combinations of useful characters in the offspring. In this way, by crossing exotics with indigenous forms, entirely new qualities can be introduced into the crops of the country. The drawbacks of hybridization are the time taken in obtaining results and the experience needed before the method can be utilized for practical purposes. Obviously, in a land of mixed varieties like India, we must first know what the country grows and what the working conditions are before a method like hybridization can be employed with effect. The possibilities of improvement by selection ought to be exhausted before hybridization is attempted. Difficult as the method of hybridization is, great advances in the improvement of the variety have been obtained by its means. In Canada and the North of the United States, a high quality bread wheat, known as Marquis, obtained by Messrs. A. P. and C. E. Saunders by crossing Red Fife with an Indian wheat (Hard Red Calcutta) has spread over millions of acres. This is the most successful hybrid yet produced. In New South Wales and other parts of Australia, the Farrer hybrids are widely grown and have been the means of raising wheat production in that country to a high standard. The method has been successfully used in India in connexion with the production of new varieties of wheat, sugar-cane, and cotton.

Whatever method of improving the crop is adopted, certain varietal characters have proved of paramount importance in India. The first of these is yielding power. It is quite useless to bring any new variety to the notice of the people unless it yields well under their conditions. Every cultivator understands the meaning of a good crop and of a variety which can be relied upon to produce a yield above the average. Once this is assured, the success of any new introduction is certain, and no difficulty in obtaining his co-operation need be feared. Even when a new variety produces a product superior to that of the country crop, this fact by itself is of little avail unless the yield is also satisfactory. Payment for yield is easy and immediate. Payment for improved quality is often a slow process. The case of cotton is a good example of this difficulty. After a variety with an improved staple has been produced and the seed-supply has been organized, it is not easy for the cultivator to obtain a greatly enhanced price for quality. The trade maintains that small lots of improved cotton are useless and that a full premium cannot be expected unless thousands of bales are available. The Agricultural Department takes the view that the best way of obtaining thousands of bales is to encourage the grower by an immediate and substantial premium. This difficulty has not yet been satisfactorily overcome, but it will have to be solved before the cultivator will give serious attention to improved quality by itself.

Besides high yield, the successful variety must possess adaptability. That is to say, it must do well over a wide range of conditions and it must also respond successfully to improved cultivation. It is an obvious advantage to all concerned to distribute as few improved kinds as possible, and the number can be materially reduced if only those possessing great adaptability are finally adopted. The indigenous varieties in India are often defective in this respect. They only do well in a very restricted area, and when the cultivation is improved they break down at once. Thus in Rohilkhand, local varieties of sugar-cane are well suited to the primitive conditions under which they are grown, but the moment they are intensively cultivated they run to excessive vegetative growth rather than to sugar production. The same applies to the country wheats, which are often useless for more intensive cultivation, and only produce, under these conditions, a mass of weak vegetation which is at once laid by even moderate storms. Really improved types of sugar-cane and wheat, while doing well under ordinary treatment, will also respond successfully to better soil conditions. This question of the adaptability of a variety is really important in India, and should be applied rigorously to all new kinds. It used to be the fashion to say that an enormous number of new varieties would have to be evolved to suit the various tracts, but this idea has not been borne out by experience. Thus the variety of tobacco known as Type 28, originally evolved at Pusa to suit the conditions in the Tirhoot Division of Bihar, has been found to do well in South Bihar, in Orissa, in Madras, in the Central Provinces and Central India, in the United Provinces and also in Burma.

Like many other successful things, the improved variety is a compromise, and does not depend on excellence in a single character. It follows that too much attention must not be paid to one character, however important it may be in itself.

Rust-resistance in wheat is a good case in point. Before the present Agricultural Department was started, a great deal was heard about the desirability of obtaining rust-resistant wheats for India. It was thought that once these could be secured, all would be well with the wheat crop. In reality, however, this is not the case. Many rust-resistant wheats in India are quite useless for any purpose beyond plant-breeding. The quality of rust-resistance is often united with so many weaknesses that the wheats are little more than curiosities. Naturally, the ideal wheat will be highly rust-resistant, but, in practice, it is better to unite with vigour, adaptability, good yielding power, good quality, and good straw, a fair degree of rust-resistance, than to pay too much attention to this one point.

The benefit obtained by the substitution of the country crop by a more efficient variety does not end with the increase in the yield per acre which results. Great as this advantage is, when we look at the total annual increment obtained over large areas, it is nevertheless a small matter compared with the enormous possibilities of an improved variety plus more intensive cultivation. It is this combination of a more efficient plant and improved methods of agriculture which holds out such great hope for the future. What these are will be evident from a few of the results obtained in the case of sugar-cane and wheat. At the Shahjahanpur Experiment Station in Rohilkhand, the potentialities of the alluvium as a producer of sugar have been explored. In spite of the short growing period, yields of country sugar as high as 100 maunds to the acre have been obtained by simultaneously improving the variety and the method of growing it. These figures are the average of nine selected varieties from 1914 to 1919 when grown on a field scale. The ordinary yield of sugar in the United Provinces is about 33 maunds to the acre, so that by the combination of more efficient varieties and better cultivation the yield can be increased threefold. Until the nitrogen problem is solved, it is not possible for every cultivator to copy these methods. The figures, however, show the possibilities which the future holds out, and the new methods point the way. The advantages of more intensive methods of cane-growing do not end with this crop. The land is left in such good condition that yields of over 36 maunds of wheat to the acre have been grown after sugar-cane as well as heavy crops of fodder and gram. The land only needs to be well manured once in four years and well cultivated afterwards to produce these results. In North Bihar, yields of over 40 maunds to the acre of Pusa 4 have been grown on a large scale under estate conditions, and similar crops have been obtained on the black soils of Central India with manure and water. In all these cases, the yield is about three times as much as a good local crop. Some years ago such results would have been considered impossible. Every year examples of such yields are increasing in number, and there is little doubt that in favoured tracts like Rohilkhand, Oudh, North Bihar, and Central India, more efficient varieties, combined with better cultivation, hold out enormous future possibilities. These areas have the advantage of a good rainfall, large supplies of subsoil water for irrigation, and a temperature very favourable for growth. Given a supply of organic matter in the right condition, moderate amounts of irrigation water, efficient varieties and good cultivation, the present production can be doubled in a few years.

## BIBLIOGRAPHY

Administration Report of the Department of Agriculture of the Punjab, 1920–1, Lahore, 1921.

Ball, C. R., and Clark, J. A.—Marquis wheat, *Farmers' Bull. 732, U.S. Dept. of Agr.*, 1916.

Buller, A. H. R.—Essays on wheat, New York, 1919.

Burt, B. C., Howard, A., and Howard, G. L. C.—Pusa 12 and Pusa 4 in the Central Circle of the United Provinces, *Bull. 122, Agr. Research Institute*, Pusa, 1921.

Clarke, G., Naib Husain, M., and Banerjee, S. C.—Notes on improved methods of cane cultivation, Allahabad, 1919.

Clouston, D.—The improvement of cotton cultivation in the Central Provinces studied from an economic point of view, *Agr. Jour. of India* (*Indian Science Congress Number*), xii, 1917, p. 29.

Guthrie, F. B.—William J. Farrer and the results of his work, *Bull. 22, New South Wales Dept. of Agriculture*, 1922.

Howard, A.—Pusa 12, *Agr. Jour. of India*, x, 1915, p. 1.

————The improvement of crop-production in India, *Jour. of the Roy. Soc. of Arts*, lxviii, 1920, p. 555.

Howard, A., and Howard, G. L. C.—Some varietal characters of Indian wheats, *Mem. of the Dept. of Agr. in India* (*Botanical Series*), ii, 1909.

————————Wheat in India, Calcutta, 1910.

## VIII

# THE DISTRIBUTION OF SEED

THE creation of an improved variety of a crop is, by itself, of no practical advantage to the country. It possesses only a potential value. The new variety must be taken up by the people, and must be welded into the rural economy before a real economic result can be achieved. This involves the distribution of the seed of the new kind to the actual cultivator. In most European countries, where the general standard of production is relatively high and where the cultivators are well to do, the introduction of new varieties is automatically carried out by private agency in the form of seedsmen. The State is not involved at any stage. In India, however, conditions are quite different. The general poverty of the people prevents the payment of the high premium for improved varieties, which makes the work of the seed-merchant both possible and profitable. In consequence, he does not exist in India, and no other indigenous agency for the distribution of pedigree seed to the cultivators has ever been developed. Methods of seed-distribution suited to the local conditions have, therefore, had to be devised by the Agricultural Department. This undertaking has been in progress for more than ten years, and sufficient experience has been obtained to make possible a review of the methods adopted and of the general results obtained.

The poverty of the average Indian cultivator not only renders the existence of the seedsman impossible, but also rules out all question of large direct profit in any scheme of seed-distribution organized by the State. Under the most favourable circumstances, the work can do little more than pay its way. The elimination of profit, however, has one great advantage. It enables the question of the maintenance of the improved variety to be considered on its merits. This is out of the question in Europe, where a distinct commercial advantage is gained by providing a constant succession of new varieties. These are needed to make the business of the seedsman a profitable undertaking. New varieties are introduced, not because existing kinds are worn out, but because fresh varieties pay the introducer. In India, the position is quite different. New varieties in this country cost time and money to produce and also to distribute, so it is an obvious advantage to all concerned to make them last as long as possible. It follows, therefore, that seed-distribution schemes should not be started until a really good variety, suitable for large tracts of country, has been obtained. A great deal of knowledge and considerable judgement are needed in deciding which variety should be adopted. Once it is obtained, every effort should be made to establish it. Patience is needed in the experiment station phase; drive in the distribution. Once a seed-distribution scheme with a good kind is launched, there is no advantage to be obtained by changing the variety unless something substantially better is obtained. To follow one variety by another, without just cause, creates confusion, minimizes public confidence in the work, and reproduces all the disadvantages of the European system, based on the seedsman and his novelties.

The maintenance of an improved variety after it is obtained is an important matter. Admixture with inferior kinds must be guarded against; cross-fertilization[1] in the field must be prevented. The greatest care must be devoted to the growing and storage of the seed at the seed-farms. If a periodical change of seed from another locality is found to be an advantage, this must be arranged for. In crops like sugar-cane and potato, which are reproduced vegetatively, frequent changes of seed are essential, otherwise degeneration is rapid. In both cases, a system of hill nurseries, after the method worked out in Java for sugar-cane, will greatly prolong the life of a variety. Why crops like sugar-cane degenerate so rapidly, when repro-

[1] Self-fertilization takes place when the male and female organs of a flower are so arranged that the pollen of each flower reaches the stigma of the same flower. Cross-fertilization occurs when the stigma of a flower becomes covered with pollen from the flower of a neighbouring plant. If the plants making up the crop are all of one type, no harm results. If, however, the crop contains a few inferior varieties, rapid degeneration results from cross-fertilization.

duced from locally-grown cuttings, is a problem which urgently calls for investigation. If the cause were known, simpler methods of prevention than those now in use might be devised.

Given a few improved varieties of the more important crops, what is the ideal system of seed-distribution to adopt? This question naturally turns on the difficulties which must be overcome in the maintenance of the variety. The greatest of these is natural cross-fertilization, which occurs to very varying extents in the different Indian crops and which rapidly leads to deterioration. In wheat, rice, and jute, for example, self-fertilization is the rule, crossing is rare, and the variety can be maintained intact for long periods. In crops like cotton and tobacco, natural cross-fertilization is very common, and, if permitted, rapidly undoes the work of the plant improver. It will be convenient, therefore, to divide the crops into two classes—self-fertilized and cross-fertilized crops—and to review under each group the various methods of seed-distribution which have been tried in this country.

### *Self-fertilized crops*

In the case of the crops in which self-fertilization is the rule, two very different systems of seed-distribution—the first intensive, the second extensive—have been tried in India. The intensive method of distribution aims at the systematic replacement of the country crop by the improved variety and the consolidation of the area under a single type. The object of the extensive method is to spread the available seed as widely as possible and to induce the cultivator to work up his own seed supply. This system has been tried in Bengal with jute and rice. The seed was divided into small packets, one of which was given to a cultivator. In this way it was hoped to overcome the difficulties involved in the growing, handling, and sale of the large quantities of seed required by any intensive method of distribution. At first the extensive method promised well, but it proved impracticable and has been given up.

The methods adopted in the United Provinces, in the spread of Pusa wheats, may be taken as an example of the intensive method of seed-distribution in the case of a self-fertilized crop. After a long series of preliminary trials, two beardless varieties with white grain—Pusa 4 and Pusa 12—were finally selected as suitable for replacing the country crop. The former was adopted for the canal areas of Bundelkhand, the latter for the alluvium. Both these varieties are easily distinguished from each other and from the country crop, which consists for the most part of a mixture of bearded wheats. This circumstance has proved of the greatest use in maintaining the purity of the new varieties on the seed-farms and in checking the work of the subordinate staff working in the Districts. Besides yielding well under cultivators' conditions, these two varieties produce grain of much better quality than the country crop, a fact which was at once recognized by the people themselves in the form of a premium for the seed which varies from six to ten annas a maund (82 lb.). The first condition in carrying out a system of seed-distribution in which the country crop is systematically replaced, village by village, is a large volume of pure seed, under the direct control of the Agricultural Department. To obtain this, large seed-farms have been established in the United Provinces, which serve as the basis of the organization in the Districts. The pedigree seed grown on these central farms is used to start and re-stock a multitude of seed-farms, managed by the local notables, such as the Taluqdars of Oudh, the large cultivators, Court of Wards' Estates and District Boards. To a certain extent, these local farms are controlled by the Agricultural Department. In this way, a large quantity of pure seed is made available every year for distribution. The actual distribution of the seed to the cultivators is carried out by various local agencies, by means of seed dépôts, and also by the Agricultural Department itself. At first it was hoped that the Co-operative movement would be the best means of establishing a network of seed-stores all over the Province, from which the members could draw their supplies for sowing. Two difficulties, however, were met with in practice. In some of the best agricultural tracts, Co-operative Societies have not yet been established. Where Societies already exist, they hesitate to take up the purchase and storage of seed on anything like the scale necessary to ensure success. A modified organization—the Agricultural Supply Society—which forms a connecting link between the Central Banks and the primary societies, is now being built up for dealing with the seed-supply. If this proves successful in practice, the Co-operative movement will play an important part in all future seed-distribution schemes in India.[1] In the introduction of Pusa 4 on the areas commanded by canals in Bundelkhand, valuable assistance has been given by the Irrigation Department. In the United Provinces, the Agricultural Loans Act enables the cultivator to obtain credit for the purchase of improved varieties of seed, as well as to meet calamities due to the season. The seed is provided by the Agricultural Department, while the numerous applications for loans and most of the actual distribution are dealt with by the Irrigation Department. This arrangement ensures the placing of seed in the villages where canal water is available, and has been the means of establishing a new variety in one of the most backward tracts of the Province. In addition, several of the Court of Wards' estates have been of great use in seed-distribution. One example will illustrate the kind of work accomplished by this agency. For several years, the Katesar Estate in the District of Sitapur has actively taken up the distribution of Pusa 12. The estate possesses its own seed store, capable of dealing with 4,000 maunds of seed every year, as well as a seed-farm of 200 acres for the production of seed of improved varieties of wheat and of sugar-cane. Besides assisting these various seed-distributing agencies, the staff of the Agricultural Department is constantly engaged in introducing improved seed, by means of village demonstrations. In the Central Circle in 1920, the number of villages dealt with was 915 and the area 18,062 acres. Samples of the seed actually distributed are grown on the Cawnpore farm, and the degree of purity noted. This constitutes an effective check on the District staff. In 1920, over 50,000 bushels of seed wheat was handled by the Agricultural Department alone. These efforts to improve the wheat crop of the United Provinces, which have been developed largely by Burt and Sharma, have been in operation for nearly ten years and are now beginning to bear

[1] At the present time, the possibility of combining the District work, at present carried out by the Agricultural and Co-operative Departments, so that it can be conducted by *one* organization instead of two independent ones, is being discussed in India.

fruit. The area under Pusa 12 and Pusa 4 is now upwards of half a million acres and is rapidly extending. These varieties are now well established in popular favour, and there is a general demand on the part of the people that more money should be spent on the organization of the seed-supply. In 1922, a resolution was moved in the Council and cordially accepted by Government, that the eventual aim should be the establishment of a seed-dépôt in every *tahsil* of the Province. The Agricultural Department is steadily increasing the number of these dépôts, and in this way a solid framework is being created, by which better varieties of wheat and other crops will be made available to the cultivators all over the Province.

### *Cross-fertilized crops*

The successful distribution of the seed of crops which readily cross-fertilize in the field is attended by far greater difficulties than have to be surmounted in wheat and rice. Among this group, most attention has so far been paid to cotton. In this crop, improvement has taken place in two different directions—the first based on the increase in the yield of short-staple types, the second on the improvement of the length and quality of the fibre.

The most important seed-distribution scheme for short-staple cotton is that worked out by Clouston in Berar, where it was found that the most profitable cotton to grow, in the brief period available for the growth of this crop, was a variety known as *Roseum*, which ripened off sharply by the end of the season and which possessed a high ginning percentage. The distribution of seed was based on the seed-farm at Akola; a large portion of the work in the Districts was carried out by means of a series of seed-unions. Each seed-union consists of ten or more members and possesses a central seed-farm of 25 to 100 acres, which is supplied with selected seed from the Akola farm. Care is taken to maintain the purity of the seed of the union central farms, which is ginned separately and distributed to the members of the union; the rest being sold to other cultivators. By concentrating these seed-unions, it has been possible for the Agricultural Department to build up a large organization for the spread of *Roseum* at a comparatively small cost. The amount of seed distributed in 1917 amounted to two million pounds, when the total area under this type was estimated at 700,000 acres. The average increased profit due to the cultivation of *Roseum* is about fifteen rupees an acre. This works out at over a crore of rupees a year—a sum which in 1917 covered the cost of the local Agricultural Department twenty times over and the cost of all the Agricultural Departments in India, including Burma, about twice over.

The establishment and maintenance of a long-staple cotton is a difficult undertaking, as the fibre rapidly deteriorates unless special precautions are taken. For the last fifty years, numerous attempts have been made to establish long-staple cotton in India. The early efforts all failed. In recent years, more systematic attempts have been made, two of which will now be considered. The most serious attempt to introduce and maintain a long-staple cotton in India is that now in operation in the Tinnevelly tract, where Sampson has devoted a number of years to the establishment of two types of selected *karunganni*—Company No. 2 and Company No. 3—in a well-defined region, where the area under cotton is about 250,000 acres. As in the case of Pusa wheat in the United Provinces, the aim of the undertaking is the complete replacement of the ordinary crop by the new types. The difficulties in such work in the case of a long-staple cotton are great. The new variety must be properly fixed, its suitability for general cultivation must be proved by numerous trials extending over a number of years, it must be maintained pure on the seed-farms, crossing with inferior types must be guarded against, mixing with other seed at ginneries must be prevented, there must be an adequate supply of pure seed, the cultivators must be assisted in obtaining a fair price for quality, the methods of seed-distribution must be such as to fit in with local conditions, and, most important of all, there must be a continuous and long-sustained effort to establish and maintain pure a large and increasing island of one type, producing sufficient cotton to keep up the interest of the trade. Unless all this is done efficiently, deterioration in the quality of the lint is inevitable. The area under seed-farms in the Tinnevelly tract, at which the two varieties are maintained pure, is 1,700 acres. A system of seed-unions, on the Berar pattern, forms a connecting link between these seed-farms and the cultivators. The active assistance of the trade has helped to solve the ginning problem, and the question of an enhanced price for purity and quality. About three-quarters of the total area of 250,000 acres has already been covered by the two improved varieties, and, if the effort is maintained for the next few years, there appears to be every promise that a real improvement in Indian cotton will be achieved. Two difficulties have arisen in this work which are of more than local interest. The first was the appearance of a robust, short-stapled variety, known as *pulichai*, with a high ginning percentage, which was used for adulterating the long-staple cotton. This, if allowed to spread naturally, would have ruined the seed-distribution scheme and would soon have destroyed the reputation of the tract as a producer of long-staple cotton. The intervention of the Government, backed by the trade and the cultivators, made the growth of *pulichai* a punishable offence, and drastic measures were taken to eradicate it. These proved successful, and the seed-distribution scheme was saved. The second difficulty arose out of the uncertain distribution of the rainfall, which made it inadvisable to distribute a single variety. Company No. 2 suits a short growing season; Company No. 3 a long one. Both cottons have a similar fibre, but they have to be grown separately on the seed-farms. The demand for the one or the other varies with the season. It is possible that further investigation will remove this difficulty and a single variety will be found for the whole tract. It will be very difficult, in practice, to prevent crossing between Company No. 2 and Company No. 3, which might lead to deterioration although the fibre of both is similar.

Another recent attempt to introduce long-staple cotton has been made on the Canal Colonies of the Punjab, the variety distributed being known as 4 F. At first the results were promising, but during the last few years the lint has deteriorated and the trade is no longer so willing to pay the premium this variety at first commanded. The failure of this seed-distribution scheme appears to be due in part to the methods adopted and to the small volume of seed produced directly by the Agricultural Department. The scheme was based on an area of ten acres of selected cotton on the Lyallpur farm, the seed of which was multiplied further on certain private estates in the Canal Colonies. Besides the small area of the improved cotton on the Government farms, there was another weak point in the scheme of distribution. The seed from the best unginned cotton offered at the Government auctions was bought in, ginned, and distributed to the cultivators at a premium (above the price of ordinary seed) through the village *bannias*. Steps have been taken by the Central Cotton Committee to provide a special staff for cotton research in the Punjab and to investigate the influence of the environment on the fibre of American cotton.

These examples, taken from the seed-distribution schemes now in progress, are sufficient to indicate the nature of the work which has to be accomplished. When the size of the country and the smallness of the means of the Agricultural Department are compared, it must be conceded that the results obtained, in spite of failure in certain cases, are of the greatest promise for the future. Well-equipped seed-farms now exist in many parts of the country, the co-operation of the people and of the trade is being secured, the local staff is being trained, and a promising beginning has been made in several localities in replacing the inferior country crop by better varieties. As was to be expected, a number of weaknesses have disclosed themselves which it will be easy to avoid in future work. One of the greatest dangers is to adopt too ambitious a scheme of seed-distribution without first of all organizing large seed-farms and a proper system of seed-control. Another weakness is the temptation to bring out too many improved varieties. It must never be forgotten that only a very few improved kinds can possibly be handled by the Agricultural Department and that no scheme of seed-distribution can achieve any permanent result if it cannot be persisted in for at least ten years and the whole resources of the Department devoted to the undertaking. Work of this kind should be judged not only by the area covered but also by the progress made in systematic replacement and by the purity of the crop as grown by the people.

Two problems connected with seed-distribution have not yet been completely solved. The first is the sale of quality in the case of crops like cotton. This matter is discussed at considerable length in the recent *Report of the Indian Cotton Committee.* The second problem is the improvement of seed-storage by the individual cultivator. If this could be accomplished, the benefits of seed-distribution schemes would be advanced a stage and a really permanent improvement effected in the rural economy of India. Clearly the cultivator himself must assume a portion of the burden involved in the seed-supply, otherwise official distribution agencies are almost certain to be abused.

## BIBLIOGRAPHY

Burt, B. C., Howard, A., and Howard, G. L. C.—Pusa 12 and Pusa 4 in the Central Circle of the United Provinces, *Bull. 122, Agr. Research Institute, Pusa*, 1921.

Clouston, D.—Cotton improvement in Berar, *Agr. Jour. of India*, x, 1915, p. 148.

——————The improvement of cotton cultivation in the Central Provinces studied from an economic point of view, *Agr. Jour. of India* (*Indian Science Congress Number*), xii, 1917, p. 29.

Couchman, M. E.—The agricultural classes in Madras, and how to approach them, *Agr. Jour. of India*, iv, 1907, p. 43.

——————Difficulties in the improvement of Indian agriculture, *Agr. Jour. of India*, vi, 1911, p. 282.

Evans, G.—The organization of seed farms in the Central Provinces, *Agr. Jour. of India*, viii, 1913, p. 267.

——————The improvement of the seed supply in the Central Provinces, *Agr. Jour. of India*, xii, 1917, p. 58.

Finlow, R. S.—Historical note on experiments with jute in Bengal, *Agr. Jour. of India*, xvi, 1921, p. 265.

Hailey, H. R. C.—Private farms in Oudh, *Agr. Jour. of India*, xii, 1917, p. 519.

Howard, A., and Howard, G. L. C.—The production and maintenance of pure seed of improved varieties of crops in India, *Agr. Jour. of India*, vii, 1912, p. 167.

Howard, A., Howard, G. L. C., and Abdur Rahman Khan.—The economic significance of natural cross-fertilization in India, *Mem. of the Dept. of Agr. in India* (*Botanical Series*), iii, 1910, p. 281.

——————Studies in the pollination of Indian crops, I, *Mem. of the Dept. of Agr. in India* (*Botanical Series*), x, 1920, p. 195.

Leake, H. M.—The breeding of improved cottons in the United Provinces, *Agr. Jour. of India*, x, 1915, p. 111.

Mann, H. H.—The introduction of improvements into Indian agriculture, *Agr. Jour. of India*, v, 1910, p. 6.

——————Local bodies as agents in agricultural improvement, *Agr. Jour. of India*, vii, 1912, p. 280.

Moreland, W. H.—The Oudh seed dépôts, *Agr. Jour. of India*, ii, 1907, p. 217.

Sampson, H. C.—Recent history of the cotton improvement work in Tinnevelly and Ramnad Districts, *Agr. Jour. of India*, x, 1915, p. 137.

Report of the Indian Cotton Committee, Calcutta, 1919.

Report of the administration of the Department of Agriculture of the United Provinces of Agra and Oudh for the year ending 30th June 1921, Allahabad, 1921.

Roberts, W.—American cotton in the Punjab, *Agr. Jour. of India*, x, 1915, p. 343.

# IX

## SOME ASPECTS OF THE FODDER PROBLEM

THE development of agriculture in India demands three conditions—water, combined nitrogen, and better cultivation. The supply of water and of combined nitrogen has already been considered. Improved cultivation is possible only if the existing cattle are better fed. For this purpose, the fodder supply must be increased so that the power needed for better cultivation will be available. The whole of the cattle problem of India is intimately bound up with the provision of more fodder and better fodder, obtained by methods within the means of the cultivator. A more nutritious and ample ration than that now available is the first condition of success in all the schemes of cattle improvement which have been advocated. The problem is mainly economic in character. The existing state of things follows naturally from the pressure of population—human and bovine—on the areas now under cultivation. It is only when the holdings of the cultivator produce more than is sufficient for himself and his family that the cattle will be properly fed. Any improvement in crop-production, therefore, will benefit the cattle by increasing the amount of fodder available. In addition to these indirect methods, a great deal of work is needed in the selection of more efficient varieties of fodder plants and in the discovery of more intensive methods of cultivation by which the largest possible yield can be raised from a small area. Better fodder is also necessary to increase the supply of milk in the cities and towns.

Countries like India, which have no good pastures, ought to pay particular attention to the fodder supply. There is little prospect of improving pasturage to any great extent. The rainfall is concentrated into a short period of the year, the soils are not suitable for meadow land, and the temperature of the hot season is too high for growth. Further, the pressure of population will always prevent any extension of the existing areas under grass. In spite of this disadvantage of poor pastures, the organization of reserve fodder in India, for carrying the stock over periods when no grazing at all exists, is surprisingly poor. There are no root crops and no good clover or meadow hay. The existence of the Indian ox is one of short commons with interludes of abundance. Efficiency suffers under the irregular feeding. The need of supplementing existing supplies is widely recognized, and various possible sources are being explored. In the Central Provinces, a species of bamboo is being grown for fodder purposes. In the Deccan, a simple method by which the prickly pear can be used for cattle food during periods of famine has been worked out at Poona. In Bombay, interesting investigations on the best way of improving the grasses of the uncultivated slopes are in progress, while in the Agra Division, one of the hot weather weeds of cultivation known as *bisurai* (*Pluchea lanceolata*, Oliver and Hiern) is being utilized for fodder purposes. New fodder grasses, such as Elephant grass (*Pennisetum purpureum*, Schum.), are being introduced into Southern India. Another recently introduced African species—*Pennisetum clandestinum*, Chiov.—is doing well in the North-West Frontier Province.

Unfavourable as the fodder supply is when looked at from the standpoint of the Occident, yet there are compensations. India possesses two valuable assets which only need utilization—intense solar energy and a temperature suitable for growth except in the hottest months. It must be remembered that in the production of fodder, vegetative growth only is required. Given a suitable temperature and favourable soil conditions, vegetative growth is exceedingly rapid and abundant in comparison with the development of the same crop when grown for seed. How can the intense solar energy and the favourable temperature of India be best utilized for fodder production? It is suggested that this can be done by intensive cultivation by which the largest possible yield can be obtained from a small area. For this purpose three things are necessary —water, organic matter in the right condition, and a suitable variety. Each fodder crop will have to be studied separately, and the conditions necessary for really high yields will have to be discovered. This information will then have to be utilized in working out simple methods of intensive agriculture within the means of the cultivator. That this is one of the directions of progress has been proved by recent investigations at Pusa in the case of lucerne, maize, and millets. Areas of poor land are now under intensive cultivation and are producing annually large crops of excellent fodder. Heavy manuring is necessary to begin with to bring the land into a condition suitable for rapid root-development and for making the best use of irrigation water. The water is applied by furrows wherever possible, so as to maintain the tilth, and the surface is top-dressed with leaf compost, made on the Chinese principle from leaves, cow-dung, and earth. In this way, growth is exceedingly rapid, the plants are constantly fed by the rich nitrous earth applied to the irrigation furrows, and one crop succeeds another with great rapidity. On the black soils of the Deccan at Manjri, similar results are being obtained. The high price of land and the cost of transporting fodder point to the necessity of intensive methods of agriculture for feeding the buffaloes and cows needed for the milk supply of the cities and towns of India. Only the merest beginning has been made in the intensive cultivation of fodder crops in India. Much remains to be done to perfect the details and to work out the economics of the subject, including the water required and the amount of organic matter necessary. The optimum conditions of growth for each crop will have to be discovered. Sufficient, however, has been done to indicate one of the directions of progress. Fodder production seems likely to develop from ordinary agriculture on lines similar to those by which market gardening has advanced. It will probably become a specialized branch, and split off from ordinary farming. The opportunity for immediate development is at hand in the zone surrounding large cities, where the demand for first-class fodder for transport and for dairy cattle is already great. The market is assured, water can often be obtained by means of the tube-well, and supplies of organic matter are available on the spot. All the conditions necessary for setting in motion intensive farming after the manner of China and Japan are satisfied. In these countries, which possess climatic advantages very like those of India, the yield is far superior to anything seen in this country. The fact that land near the cities of India is expensive will act as a powerful stimulus in making the most of it and in obtaining the largest possible yield.

Although India is poor in good pasture plants, suitable fodder crops abound. Maize and millets take the place of the grasses of the West ; lucerne, berseem, Persian clover (*shaftal*) —*Trifolium resupinatum* L.—and the cluster bean (*guar*)—*Cyamopsis psoralioides* DC.—represent the clover family. Properly cultivated, these crops yield enormous quantities of fodder. In a preliminary experiment at Quetta, a crop of Persian clover, following maize as a foster crop, gave over thirty-three tons to the acre. Lucerne, properly cultivated in the plains, will yield at least as much. Once fodder crops are intensively cultivated round the cities and towns, it is probable that the practice will spread slowly to rural India. Here and there well-to-do cultivators are certain to adopt the new methods and to raise their fodder supplies from a small area.

The cultivation of leguminous fodders like lucerne, berseem, and Persian clover in India has already brought to light a number of minor problems. As would be expected from experience in Europe and America, the breed of lucerne is of importance. The type grown near Kandahar, which at present supplies a good deal of the seed in India, is not very well adapted to the plains of India, as it is a hot weather variety and proves to be a slow grower during the cold season of the plains. The resulting crop is generally infested by a parasite, known as dodder, which spreads rapidly, lowering the yield and spoiling the fodder, especially when hay is desired. Comparative trials at Pusa show that the acclimatized type grown at the Remount Dépôt at Saharanpur and freshly imported seed from the Hunter Valley in Australia are far superior to Kandahar lucerne. The best of these mixtures might easily be improved by selection. Another difficulty is to make some of these leguminous fodder crops set seed in India. This failure to seed freely appears to be due to high temperature preventing the germination of the pollen grains. Seed will, therefore, have to be raised in the cooler parts of the country. Unless a reliable indigenous seed-supply is available, crops like lucerne and berseem are not likely to be taken up by the people. In addition to the seed-supply, there is another difficulty in the growth and introduction of leguminous fodders, namely, an adequate supply of nodule organisms in the soil. At first, the production of root-nodules is apt to be small and the yields poor. A rapid improvement, however, often takes place after two or three years, and then only is it possible to draw conclusions as to the suitability of the species. The results of the first few years are often quite misleading. Time and patience are therefore important in the trial of leguminous crops in new localities.

Besides intensive cultivation near cities, there are other openings for the growth of lucerne and berseem. In the Canal Colonies of the Punjab, a leguminous rotation is greatly needed on the wheat lands. In a previous chapter on irrigation and water-saving (p. 28), it was pointed out that a large proportion of the water given to the wheat crop is wasted and could easily be saved. Most of this water could be profitably used in the growth of lucerne, berseem, and Persian clover. These fodders give the best yield if the land is well manured. If, therefore, the cultivator were to put say 10 per cent. of his land into these fodders every year and were to concentrate his supply of manure on this area, this valuable rotation would rapidly raise the productive power of his holding and would also indirectly increase the duty of water. From the point of view of irrigation, these innovations would be a great advantage, as they would reduce the demand for water during February and March, when the supplies are often low. Besides the Canal Colonies, the cultivation of lucerne would be an advantage in Central India, where a great deal of the transport of produce is done by road. A portion of the crop, if made into hay, would be of great use in improving the feeding value of the grass which is harvested after the rains.

It is not sufficient merely to grow fodder crops and to feed them green. Storage is necessary for the hot weather and for the beginning of the cold season. Silos are proving very suitable for preserving maize and millets, and ensilage is taking the place of the roots of Western countries. Silage, however, can only be used locally, and is not adapted for transport. In India, straw and dried grass take the place of hay. It is here that improvement is desirable, as these fodders are of low feeding value and do little more than maintain the animals. India has no leguminous hay at present, but there is no reason why this state of things should continue. Lucerne and berseem make excellent hay, provided care is taken to preserve the leaves (the most nutritious portion) during the curing process. This can be done by completing the drying in small heaps and by taking care to stack the finished product in the early morning before it gets too brittle. The difficulty in India is not so much to get the fodder dry, but to preserve sufficient moisture during the curing process so that there is little loss of leaf in handling. Even in the very dry climate of Baluchistan, where soon after sunrise lucerne hay is much too brittle to handle, it is possible not only to make lucerne and *shaftal* hay without loss of leaf, but also to compress it to the army standard—90 cubic feet to the ton. The baling difficulties are overcome by moistening and covering the heaps of fodder for about twelve hours before the bales are made. The pressed fodder keeps perfectly for years, even when brought into such a damp climate as that of Bihar. Indeed, lucerne hay does not spoil at Pusa when stored loose. For army purposes, where supervision is possible, perhaps the easiest method of baling lucerne, berseem, and *shaftal* hay is in the form of chaff mixed with the desired proportion of broken straw (preferably one part of hay to two parts by weight of straw). The mixture is readily compressed, and if the proper amount of grain is added at the same time, a complete ration, very suitable for fodder reserves and for active service, could be produced. Such bales would greatly simplify the feeding of army animals on active service. The suitability of lucerne and *shaftal* hay for army transport purposes, under service conditions, has been proved by trials extending over four years in the Fourth Division at Quetta during the years 1915 to 1918. A saving in weight of as much as 40 per cent. was obtained, while the cost of feeding was slightly reduced. The initial difficulties have been overcome, and interesting developments should result. The great difficulty in getting these new fodders taken up in the North-West is to secure a certain market at a fair price during the early stages. It is here that the Army can prove its usefulness in the agricultural development of the Punjab. It is always a common reproach that expenditure for military purposes is entirely in the nature of an insurance, and that no direct good to the country ever results. Should, however, the Army succeed in getting the cultivators in the North-West to produce good leguminous hay, they will be able to point to a valuable piece of development work. The growth of these fodders in the rotation is essential for increasing the fertility of the soil. At the same time, one of the great transport problems—both military and civil—of Northern India will have been solved, and the way will then be clear to a better local organization of fodder. The Irrigation Department has already rendered useful help in this work by altering the existing rules so that the water rates charged for *shaftal* are fair to both sides. The interest and assistance of the Agricultural and Co-operative Departments are now needed to carry the work to new centres, so that a large volume of improved fodder will be available both for military and also for civil needs.

## BIBLIOGRAPHY

Branford, R.—Cattle stock and fodder famine in Hissar, *Agr. Jour. of India*, xvii, 1922, p. 212.

Branford, R., and Sewell, E.—Feeding experiments at the Government cattle farm, Hissar, *Bull. 130, Agr. Research Institute, Pusa*, 1922.

Brown, W. R.—Kikuyu (*Pennisetum clandestinum*): a new pasture grass for India, *Agr. Jour. of India*, xix, 1924, p. 32.

Bruce, J.—An account of some experiments in lucerne cultivation at Saharanpur Remount Dépôt, *Agr. Jour. of India*, xiii, 1918, p. 254.

Burt, B. C.—Some experiments in manuring for double cropping, *Agr. Jour. of India*, xii, 1917, p. 536.

Clouston, D., and Plymen, F. J.—Principal fodders in the Central Provinces and Berar including the small bamboo (*Dendrocalamus strictus*), *Agr. Jour. of India*, xv, 1920, p. 380.

Henderson, G. S.—Berseem as a new fodder crop for India, *Bull. 66, Agr. Research Institute, Pusa*, 1916.

Howard, A.—The improvement of fodder production in India, *Jour. of Dairying and Dairy Farming in India*, iii, 1916, p. 189. (Reprinted in the *Agr. Jour. of India*, xi, 1916, p. 391.)

————A note on the baling of shaftal and lucerne hay for army transport, *Jour. of Dairying and Dairy Farming in India*, v, 1917, p. 61. (Reprinted in the *Agr. Jour. of India*, xiii, 1918, p. 717.)

Howard, A.—An improved method of lucerne cultivation, *Agr. Jour. of India*, xix, 1924, p. 276.

Howard, A., and Howard, G. L. C.—Clover and Clover Hay, *Agr. Jour. of India*, ix, 1916, p. 71.

————————Leguminous crops in desert agriculture, *Agr. Jour. of India*, xii, 1917, p. 27.

————————The agricultural development of North-West India, *Agr. Jour. of India* (*Indian Science Congress Number*), xii, 1917, p. 67.

————————The saving of irrigation water in wheat growing, *Bull. 118, Agr. Research Institute, Pusa*, 1921.

Kikuyu grass (*Pennisetum clandestinum* Chiov.), *Kew Bulletin*, 1921, p. 85.

Knight, J. B.—Prickly pear as a cattle food, *Bull. 97, Dept. of Agr., Bombay*, 1920.

Mann, H. H.—Fodder crops of Western India, *Bull. 100, Dept. of Agr., Bombay*, 1921.

Matson, J.—Lucerne—why an irrigated crop, *Jour. of Dairying and Dairy Farming in India*, vi, 1918, p. 7.

Parr, A. E.—A valuable cattle fodder, *Agr. Jour. of India*, xvi, 1921, p. 206.

Report of the Agr. Research Institute, Pusa, 1920–1, p. 18.

Sen, J. N.—The composition of some Indian feeding stuffs, *Bull. 70, Agr. Research Institute, Pusa*, 1917.

The improvement of fodder and forage in India (Papers read before a joint meeting of the Sections of Agriculture and Botany, Indian Science Congress, 1923), *Bull. 150, Agr. Research Institute, Pusa*, 1923.

# X

# COTTON

A STUDY of the distribution of the cotton crop in India shows that it is confined to the drier regions. Cotton either does not occur at all or is of very minor importance in Oudh, Bengal, the alluvial tracts of Bihar and Assam, and the rice areas of the Central Provinces. In Eastern India and Burma the crop is only found in the hill tracts or where the permeability of the soil is above the average. About half the 20,000,000 acres under cotton in India produces the class of fibre known in the trade as Oomras. This grade comes from a compact black soil area in the middle of Peninsular India which includes the Central Provinces, the States of Central India, the Bombay Deccan, and a large portion of Hyderabad. The greatest concentration of cotton is in the District of Amraoti in the centre of the Oomras tract, where over 50 per cent. of the cropped area consists of cotton. Only one-seventh of the total acreage occurs on the alluvium of the Indo-Gangetic plain, the chief centres of cultivation being found on the canal-irrigated areas of the Punjab and Sind, in the Province of Agra, and in certain of the Rajputana States. We should expect from the distribution of cotton that one of the chief requirements of this crop would be a permeable, well-aerated soil. This is confirmed by the results of various experiments in lysimeters at Pusa, where it was found that any diminution of the aeration of the roots checked the growth, while an increase in the permeability of the soil was followed by more rapid development.

The problem of improving the yield and quality of Indian cotton is a many-sided one, and is therefore of considerable scientific interest. The yield of lint depends on a number of factors such as the general vigour of the plant, the mode of branching, and the percentage of fibre extracted in the process of ginning. The crop is sown in the early rains, but often has to complete its activities when the monsoon is over and the temperature has fallen considerably. The seedling stage is often a difficult one, so the young plants must possess hardiness and must develop quickly. The root-system of an efficient variety must therefore possess great adaptability. The value of the produce depends not only on the total weight of lint, but also on the length, strength, uniformity, colour, and fineness of the fibre. Some of these desirable characters, such as length of fibre, appear to be incompatible with a short growing period and with a high ginning percentage. Both the length and strength of the fibre are greatly influenced by the environment, while the ravages of pests can always be relied upon to reduce the yield.

The average yield of cotton in India is exceedingly low, and works out at about a maund (82 lb.) of lint to the acre. This is only one-third of the average production of the cotton areas of the United States. Most of the crop is raised on the natural rainfall, and, outside the plains, very little is irrigated either from wells or canals. Perhaps one cause of the low production in Peninsular India is the frequent occurrence of the cotton-millet rotation and the small part played by leguminous crops in the general economy of cotton. This rotation is an unfavourable one from the point of view of combined nitrogen, as both the cotton stalks and the cow-dung produced from the fodder crop are mostly burnt for fuel. In addition, there is little nitrogen given back to the soil by direct manuring or by the growth of green manure. Such a rotation must, therefore, involve a steady loss of nitrogen. Nevertheless, the fertility of the soil does not appear to diminish. Everything therefore points to a relatively large amount of direct fixation of atmospheric nitrogen by the various soil organisms to make up for the various losses. Can anything be done to improve this state of things? It is more than probable that a comparatively small increase in the content of organic matter in the cotton soils of Peninsular India, combined with simple improvements in soil management, would not only greatly stimulate nitrification, but would also speed up the amount of direct fixation from the air. This would lead to more vigorous growth and to an increased production of lint. It appears to be the most promising direction of increasing the yield of cotton in the tract producing Oomras, where the average out-turn is lower than in many of the other cotton-growing areas of the country.

In the days of the East India Company and since the Mutiny, a great deal of attention has been paid to the improvement of Indian cotton. One of the most serious attempts was that made during the war between the Federals and Confederates, when the supply of American cotton for the Lancashire mills fell off and alternative sources of raw material were explored. The seed of long-staple cotton from the United States was extensively introduced into the Central Provinces, Bombay, and other parts of India. With the exception of a small area round Dharwar and an admixture of American types in the general crop in the north of the Central Provinces, very few traces of these early efforts to improve the staple by direct introduction were in existence when the problem was examined anew by the present Agricultural Department in the early years of this century. In a recent paper, dealing with these early efforts, Clouston states that 'the lint of exotic cottons grown in the Central Provinces invariably deteriorates after a certain number of years. Our Upland Georgian has a distinctly weak fibre, due to its having been cultivated for about half a century under the unfavourable soil and climatic conditions which obtain in these parts'. Very similar results have followed most of the trials of exotics carried out during the present century. When the amount of time and energy expended on the introduction of varieties from overseas is considered, it is not likely that the problem of improving cotton-production in India will be solved by further efforts of the same kind.

After the foundation of the present Agricultural Department by Lord Curzon in 1904, cotton was one of the crops studied, and an enormous amount of work has been carried out in practically all the important areas during the last fifteen years. In many cases, as in Clouston's investigations on *Roseum* cotton in Berar, Sampson's work on *karunganni* cotton in Tinnevelly, Parr's researches on Aligarh white-flowered cotton in the Province of Agra, the work on American cotton in the Punjab (Milne) and United Provinces (Burt), large schemes of seed-distribution have arisen which have affected many thousands of acres. Besides these a number of other investigations—of Hilson in Madras and Kottur and Patel in the Bombay Presidency—are beginning to influence the cultivators. The results of this work have been so recently dealt with in the *Report of the Indian Cotton Committee* and in its appendices that it is unnecessary to go over the same ground again. One feature of these investigations, however, calls for notice, namely, the local character of a large portion of the work. This was inevitable from the organization of the investigations on Provincial lines and from the necessity of practical results in the shape of improved varieties for seed-distribution, so that the utility of the Agricultural Department could be established. The investigators were not altogether free in the selection of problems for investigation. In spite of these disadvantages, a certain amount of fundamental research work has been accomplished in addition to a mass of results of local interest. Leake has published a good deal of work on the inheritance of characters in cotton, while Clouston and Sampson have clearly defined the chief factors underlying cotton improvement in the Central Provinces and in the Tinnevelly tract of Madras respectively. Gammie has published a detailed classification of the species and varieties of Indian cotton which has been utilized freely by other workers. Investigations on the incidence of cross-fertilization in the various types of cotton grown in India have been carried out by many workers. These pieces of work, valuable as they are, by no means exhaust the subject. A great deal of further research is needed on the general principles underlying cotton-production, a need which has been recognized by the recently formed Central Cotton Committee. This is a statutory committee, and consists of representatives of the trade, the growers, and the Agricultural Department. One of the proposals put forward by this body and adopted by the Legislative Assembly is to finance the large amount of fundamental research needed on the production and testing of Indian cotton by means of a small cess of two annas a bale, levied on the cotton used in and exported from India. This will yield about five lakhs of rupees a year.

There are several directions in which the cotton plant in India needs detailed investigation. Some of these will now be considered. They fall into two groups—the first is concerned with the variety, the second with the environment.

### *Varietal Factors*

The first condition of progress in the improvement of Indian cotton is a botanical survey of the indigenous crop and the isolation of the various unit species which make up the existing varieties. This means the ultimate classification of Indian cotton. Except in a few cases, such as Youngman's survey of the forms grown in parts of the United Provinces, classification has not proceeded more than half-way and has only reached the point known as the botanical variety. When the composition of the varieties of Indian cotton is known, a mass of valuable material will be available for the plant-breeder and for the more fundamental research which is so necessary for future progress, including the intensive study of the unit species. An essential part of a botanical survey is the study of root-development and its relation to the soil type. The roots of the cotton plant and the soil must be suitably geared right through the growth period if the maximum yield is to be obtained. It will also be a great advantage in all new varieties if the crop ripens off sharply with the minimum number of pickings and with no great range in the quality of the lint. It is probable that root fitness will be found to be one of the chief factors involved in such questions.

Given a suitable collection of unit species, two of the fundamental problems underlying the production of long-staple cotton in India can be immediately attacked. The first is the precise relation which exists between the length of staple and high ginning percentage and the extent to which these two can be profitably combined. It is well known that, *as a rule*, long-staple cottons have a low ginning percentage, while short staple and high ginning percentage are frequently combined. It seems too much to hope, therefore, that a cotton will ever be found which combines the longest fibre and the highest ginning percentage. On the other hand, an examination of a large range of unit species might easily show that a much more effective union of staple and ginning percentage can be obtained than is found in the mixtures usually cultivated. Kottur claims that he has combined length of fibre and high ginning percentage in a cross between pure lines of *G. herbaceum* and *G. neglectum*. It will be interesting to see whether these new forms are of use in the field and yield well or whether the energy required to give long fibre and high ginning percentage simultaneously is obtained at the expense of the total yield. A second fundamental problem to be attacked is the effect of cross-pollination on the maintenance of the variety. As is well known, cross-pollination in cotton often leads to increased vegetative vigour in the following generations and also to loss of quality in the lint. In the case of long-staple cotton, the tendency for some years has been to prevent all crossing and to sacrifice the advantages of increased vigour in favour of the maintenance of the quality of the lint. Careful investigation in two directions is needed on this question. In the first place, the effect of long-continued self-fertilization in a pure line should be examined. In the second place, the possibility of the control of crossing between pure lines, so selected that there is no disturbance in the quality of the lint, needs a thorough exploration. In this way the advantages of increased vigour might be secured without any alteration in the fibre. It is by no means impossible that the plant breeders of the future will not be able to sacrifice such a great advantage as would be conferred by cross-fertilization working between known limits and producing known results.

### *Environmental Factors*

While the breed obviously exercises the greatest influence on the lint, various important modifications are introduced by the environment. The fibre is altered in such characters as length, silkiness, colour, and strength by various soil and climatic conditions. It is therefore essential, in selecting improved cottons, to find one which is affected to the least possible extent by natural variations in the environment. Unless this is provided for, the demand for a uniform product, so necessary for the trade, will not have been satisfied.

The direct relation which appears to exist in India between the length of staple and the growth period needs further investigation. If it is firmly established that tracts like Berar and the Agra Division are naturally most suited to the production of a short fibre, then the efforts of the plant breeder will naturally be directed to the increase of the yield per acre rather than to the increase of the length of the lint. The experience obtained with Cambodia cotton in Peninsular India appears to support the general impression that a connexion between the length of fibre and the duration of growth undoubtedly exists. Grown on garden lands under irrigation, Cambodia produces lint of from three-quarters to over an inch in length. Under dry cultivation the fibre shortens, and reaches only five- to seven-eighths of an inch. Should certain tracts be found suitable for short staple only, it will be a great advantage to keep the long-staple and short-staple areas separate and to keep the two grades apart in marketing. The recently enacted Cotton Transport Act will greatly assist in this work.

Just as the aeration of the soil affects growth, so, other things being equal, this factor appears to influence the length and character of the fibre. The most striking experiments on this subject are those of Clouston, which were carried out on the laterite (*bhata*) and black soils at Chandkhuri in Chhatisgarh. Here the rainfall is high, irrigation is available, and the two soil types differ markedly in permeability. The *bhata* soils are lateritic in origin and exceedingly porous. The black soils are stiff and easily become impermeable. When *Roseum* was grown side by side on these two classes of soils, the permeability of *bhata* was found to increase the length and improve the quality of the fibre to such an extent that the trade experts, to whom the two sets of samples were submitted, considered the produce of the *bhata* soil to be that of a superior variety. On the Surat farm, in the Broach area, it has recently been discovered by Mann and his staff that the limiting factor in cotton-production is poor soil-aeration. A considerable increase in yield is obtained by growing the crop on flat ridges, which improve surface-drainage during the rains, increase the aeration, and stimulate root-development. It will be interesting to find what effect this increased aeration has on the lint characters. These two results illustrate the importance of fundamental research in agricultural development. But for the investigations which were carried out some years ago in India on the importance of soil-aeration in growth, it is safe to say that the possibilities of the *bhata* soils for the production of long-staple cotton would never have been explored. Although the Surat farm has been in operation for over twenty-five years, and numerous experiments had been carried out to discover the best means of increasing the yield of cotton, no real progress was made till the possibility that poor soil-aeration was the limiting factor in growth was put to the test.

While an increase in aeration improves the lint, other factors, still undiscovered, connected with poor soil-ventilation appear to operate in the opposite direction. The rapid spread of American cotton in the Punjab and Sind has, during the last few years, been associated with a marked falling-off in the strength of the fibre. The same thing has occurred in the Central Provinces and other parts of India where American cotton has been tried. In the Central Provinces the result appears to be associated with the cracking of the soil and the destruction of the roots during the ripening period. It is not impossible that the loss of permeability of the soil in the Punjab, due to continuous irrigation, may operate in the same direction and that the explanation of the loss of strength of the fibre of American cotton in this Province will be associated with a falling-off in efficiency of the root system during the ripening period. The matter is one which calls for immediate investigation both on the alluvium and on the black soils. Weak fibre might easily prove to be a consequence of unsuitable gearing of roots and soil. Two other drawbacks in cotton-growing in India—boll-shedding and damage by insects—appear to be influenced by the environment. Investigations are in progress to discover the factors which are responsible for the shedding of bolls and which render the cotton plant so susceptible to the attacks of insects.

There is one aspect of the effect of environment on quality which is of great promise in the improvement of the staple of Indian cottons, namely, intensive cultivation. Most of the Indian cotton is a rains crop, and the yield depends on the vagaries of the monsoon. It seems possible, by intensive methods on suitable soils and with suitable varieties, to eliminate most of the limiting factors and to raise a large quantity of long-staple cotton from a small area. The experience already obtained with Cambodia cotton on certain soils in Madras, on the *bhata* areas of Chhattisgarh, and on the old poppy lands of Central India all point to almost certain success in this direction. Given the ideal soil and moisture conditions, which are easily possible by the intensive cultivation of open black soils, the question of the choice of variety enters on a new phase, and types entirely unsuitable for dry cultivation may easily prove to be the best for the new conditions. Varieties at present unknown in India, or which have failed entirely for extensive cultivation, may come into their own if we add organic matter and water to some of the soils of Peninsular India.

## BIBLIOGRAPHY

Ajrekar, S. L., and Bal, D. V.—Observations on the wilt disease of cotton in the Central Provinces, *Agr. Jour. of India*, xvi, 1921, p. 598.

Annual Reports of the Central Cotton Committee, Bombay, 1922 and 1923.

Burt, B. C.—American cotton in the Central Circle, United Provinces, *Agr. Jour. of India*, viii, 1913, p. 339.

———Cawnpore American cotton, II. Further field trials (1918–20). Spinning trials and market organization, *Bull. 126, Agr. Research Institute, Pusa*, 1922.

———A preliminary study of the demand for various types of Indian cotton, Bombay, 1922.

———The Indian Central Cotton Committee, *The Empire Cotton Growing Review*, i, 1924, p. 25.

Burt, B. C., and Nizamuddin Haider.—Cawnpore American cotton. An account of experiments in its improvement by pure line selection and of field trials, 1913–17, *Bull. 88, Agr. Research Institute, Pusa*, 1919.

———The Bundelkhand cottons. Experiments in their improvement by pure line selection, *Bull. 123, Agr. Research Institute, Pusa*, 1921.

Clouston, D.—Artificial fertilizers for cotton in the Central Provinces. *Agr. Jour. of India*, ii, 1907, p. 116, and iii, 1908, p. 245.

———Cotton cultivation in the Central Provinces and Berar studied from an economic aspect, *Agr. Jour. of India*, vi, 1911, p. 353.

———Cotton improvement in Berar, *Agr. Jour. of India*, x, 1915, p. 148.

———The improvement of cotton cultivation in the Central Provinces studied from an economic point of view. *Agr. Jour. of India (Indian Science Congress Number)*, xii, 1917, p. 29.

Coventry, B.—Note on the present position of cotton investigations in India, *Bull. 26, Agr. Research Institute, Pusa*, 1911.

Dastur, J. F.—A preliminary account of the investigation of cotton wilt in the Central Provinces and Berar, *Agr. Jour. of India*, xix, 1924, p. 251.

Fletcher, F.—The improvement of the cottons of the Bombay Presidency, *Agr. Jour. of India*, i, 1906, p. 351.

Gammie, G. A.—The Indian cottons, *Mem. of the Dept. of Agr. in India (Botanical Series)*, ii, 1907.

——————Work done towards the improvement of cotton in the Bombay Presidency, *Agr. Jour. of India*, iii, 1908, p. 135.

——————Cultivation of cotton in India, *Agr. Jour. of India*, iii, 1908, p. 164.

——————Caravonica cotton, *Agr. Jour. of India*, v, 1910, p. 249.

——————The present position and prospects of cotton cultivation in India, *Agr. Jour. of India*, v, 1910, p. 335, and vi, 1911, p. 40.

Gaskin, L. E. P.—Cotton cultivation and trade in the Central Provinces and Berar, *Agr. Jour. of India*, ii, 1907, p. 171.

Harland, S. C.—In-breeding in cotton and its importance to the plant breeder, *Agr. Jour. of India*, xviii, 1923, p. 465.

Henderson, G. S.—Exotic cottons in Sind, *Agr. Jour. of India*, vi, 1911, p. 134.

Hilson, G. R.—' Northerns ' cotton, *Agr. Jour. of India*, xiv, 1919, p. 300.

——————Cambodia cotton (*Gossypium hirsutum*): its deterioration and improvement, *Agr. Jour. of India*, xvi, 1921, p. 235.

——————Uniformity of length of cotton hairs, *Agr. Jour. of India*, xvi, 1921, p. 564.

——————Methods of examination of certain characters in cotton, *Bull. 138, Agr. Research Institute, Pusa*, 1922.

Keatinge, G. F.—Co-operative Societies for the sale of cotton in the Southern Maratha country, *Agr. Jour. of India*, xiii, 1918, p. 121.

Kottur, G. L.—Improved types of cotton for the Southern Maratha country, *Agr. Jour. of India*, xiv, 1919, p. 165.

——————Kumpta cotton and its improvement, *Mem. of the Dept. of Agr. in India (Botanical Series)*, x, 1920, p. 221.

——————Note on protecting cotton flowers from natural crossing, *Agr. Jour. of India*, xiv, 1919, p. 177.

——————Cross-fertilization and sterility in cotton, *Agr. Jour. of India*, xvi, 1921, pp. 52 and 406.

——————An improved type of cotton for the Dharwar-American tract, *Agr. Jour. of India*, xvii, 1922, p. 347.

——————Studies in inheritance in cotton, I. History of a cross between *G. herbaceum* and *G. neglectum*, *Mem. of the Dept. of Agr. in India (Botanical Series)*, xii, 1923, p. 71.

Kottur, G. L., and Patel, M. L.—Malformation of the cotton plant leading to sterility, *Agr. Jour. of India*, xv, 1920, p. 640.

Leake, H. M.—Studies in Indian cotton, *Jour. of Genetics*, i, 1911, p. 205.

——————Experimental studies in Indian cottons, *Proc. Roy. Soc.* B, lxxxiii, 1911, p. 447.

——————A preliminary note on the factors controlling the ginning per cent. of Indian cottons, *Jour. of Genetics*, iv, 1914, p. 41.

——————The breeding of improved cottons in the United Provinces, *Agr. Jour. of India*, x, 1915, p. 111.

Leake, H. M., and Parr, A. E.—The problem of the improvement of cotton in the United Provinces of Agra and Oudh, *Agr. Jour. of India*, vi, 1911, p. 1.

——————————The problem of the improvement of the indigenous cottons of the United Provinces, *Agr. Jour. of India*, viii, 1913, p. 47.

Leake, H. M., and Ram Prasad.—Notes on the incidence and effect of sterility and cross-fertilization in the Indian cottons, *Mem. of the Dept. of Agr. in India* (*Botanical Series*), iv, 1911, p. 37.

——————————Studies in Indian cottons, Part 1. The vegetative characters, *Mem. of the Dept. of Agr. in India* (*Botanical Series*), vi, 1914, p. 115.

MacKenna, J.—The early history of cotton in Bombay, *Agr. Jour. of India*, xiii, 1918, p. 389.

Main, T. F.—The improvement of cotton in Sind, *Agr. Jour. of India*, ii, 1907, p. 349.

Moreland, W. H.—Conditions determining the area sown with cotton in the United Provinces, *Agr. Jour. of India*, i, 1906, p. 37.

Patel, M. L.—Studies in Gujarat cottons, I and II, *Mem. of the Dept. of Agr. in India* (*Botanical Series*), xi, 1921, p. 75, and xii, 1924, p. 185.

Purshotamdas Thakurdas.—Co-operative Societies and marketing of cotton, *Agr. Jour. of India*, xiv, 1919, p. 168.

Ram Prasad.—Note on the probability of an inter-relation between the length of the stigma and that of the fibre in some forms of the genus *Gossypium*, *Bull. 137, Agr. Research Institute, Pusa*, 1922.

Report of the Indian Cotton Committee, Calcutta, 1919.

Roberts, W.—American cotton in the Punjab, *Agr. Jour. of India*, x, 1915, p. 343.

——————American cotton and American cotton sales in the Punjab, *Agr. Jour. of India*, xi, 1916, p. 238.

——————Some of the problems arising out of the successful introduction of American cotton in the Western Punjab, *Agr. Jour. of India*, xiv, 1919, p. 42.

Sampson, H. C.—The introduction and spread of Cambodia cotton in the Madras Presidency, *Agr. Jour. of India*, vi, 1911, p. 365.

——————Recent history of the cotton improvement work in Tinnevelly and Ramnad Districts, *Agr. Jour. of India*, x, 1915, p. 137.

Sampson, H. C.—Some aspects of cotton improvement in India, *Agr. Jour. of India*, xv, 1920, p. 277.

Thadani, K. I.—Some notes on cotton in Sind, *Agr. Jour. of India*, xv, 1920, p. 393.

Thadani, K. I.—Linkage relations in the cotton plant, *Agr. Jour. of India*, xviii, 1923, p. 572.

Thomas, R.—The improvement of Tinnevellies cotton, *Agr. Jour. of India*, xiv, 1919, p. 315.

Thompstone, E.—The ginning percentage of cotton in its relation to the season, *Agr. Jour. of India*, xv, 1920, p. 568.

Vagholkar, B. P.—Ginning percentage of cotton in its relation to season, *Agr. Jour. of India*, xvi, 1921, p. 690.

Venkayya, P.—Cambodia cotton in India, *Agr. Jour. of India*, vi, 1911, p. 160.

# XI

## WHEAT

WHEAT is one of the chief cold-weather food crops of India ; every year about 9,000,000 tons are raised from an area over 30,000,000 acres in extent. The average production is therefore extremely low, only 8·2 maunds (672 lb.) to the acre. Most of the crop is consumed by the people, and, even after good harvests, not more than 10 per cent. is exported. About three-quarters of the produce comes from the drier alluvial areas of North-West India. Unlike rice and millets—the chief monsoon food crops—wheat is a plant which does not require a great deal of moisture. For heavy yields, the supply of nitrogen must be adequate, a requirement which is most easily met in India by keeping up the content of organic matter in the soil rather than by direct manuring. Besides sufficient moisture and organic matter, temperature is an important factor in wheat production in this country. This limits the growth period in both directions. After the monsoon, the sowing of wheat is regulated by the point reached in the gradual cooling of the seed-bed. Towards harvest, the crop has to ripen under a rapidly ascending temperature when hot dry winds are frequent. At both ends, therefore, the growth period is temperature limited, a fact which not only restricts the choice of varieties to early maturing types which grow rapidly, but also influences the geographical distribution of the crop. There is a tendency in some parts of India for the cultivators to pay too little attention to the temperature factor. They often sow too early and also favour the cultivation of varieties which ripen on the late side. These tendencies are probably a result of poverty and of the temptation to strive after the maximum crop. In favourable seasons they are successful, but experience shows that, over an average of years, those varieties which ripen off well within the growth period give the best results. To some extent, therefore, wheat growing in India is a gamble in temperature. In most of the wheat areas, the indigenous varieties possess low tillering power, weak straw, and a tendency to leafy growth. The ears are usually bearded and the grain is of poor quality compared with the best produce of Canada and North America. Indian wheat, however, possesses two great advantages from the millers' point of view. The percentage of moisture is often low; the yield of flour high. Both these circumstances tend to enhance the price obtained in the Home markets. The essential requirement in an efficient variety is speed of growth. The season is short, there is practically no latitude, while both moisture and nitrogen are liable to be deficient. Varieties which cannot give a crop under unfavourable conditions are useless. These considerations explain the failure of exotics imported from Europe and America.

A good deal of work has been done on the improvement of wheat production in India since the subject was considered by the Board of Agriculture in 1906. A botanical survey of the varieties cultivated has been made in all the chief wheat-growing areas, the results of which were published in *Wheat in India* in 1909. Nearly forty botanical varieties have so far been identified. Ten of these belong to the macaroni wheats (*T. durum* Desf.), six to the group of dwarf wheats (*T. compactum* Host.), there is one variety of emmer (*T. dicoccum* Schrk.) and nineteen varieties of bread wheat (*T. vulgare* Vill.). No spelt wheats occur. The single variety of emmer is found in the south of the wheat-growing area—in Bombay, Madras, Mysore, and to a very small extent in Berar. The drought-resistant dwarf wheats are chiefly found in the Western Punjab. The chief points of interest in the distribution of varieties concern the bread and macaroni wheats. The former are typical of the alluvial soils of the plains, the latter of the dry black soil areas south of the Jumna. This separation results from differences in the soil type. On the alluvium, bread wheats form the usual secondary root-system and proceed to tiller. Fairly large yields of moderate-sized grain result. On the black soils, except in certain favoured localities, macaroni wheats are the rule. The upper soil dries so rapidly after sowing that there is hardly any development of the secondary root-system, and practically no tillering. The crop develops slowly on the primary roots; the grain is always large and the yields extremely low. Some of the macaroni wheats have black awns, in some cases with a certain amount of blackening of the chaff. Three of the varieties of common wheat in Bihar and Orissa have black chaff and black awns; while in this tract short rounded grains are common. In this region, the classification has been carried to the unit species.

Up to the year 1908, it was generally believed in the wheat trade that India could only produce wheats of relatively poor grain quality. This conclusion was based on a long practical experience of the wheats exported from the country. In order to ascertain whether this necessarily applied to all Indian varieties, a large number of the types isolated at Pusa were sent to England in 1908 and subsequent years for complete milling and baking tests, the work being undertaken by Dr. A. E. Humphries, a former President of the Incorporated National Association of British and Irish Millers. At the same time, samples of the varieties commonly grown for export in Northern India and on the soils of the Peninsula were included for purposes of comparison with the new types. These trials proved conclusively (1) that many of the new varieties possess the character of free-milling and also yield flour and loaves of the same class as that produced from the strongest North American grades, and (2) that strong, free-milling wheats can be grown at Pusa and that such varieties are more suitable for export than the wheats ordinarily shipped from India.

The next step was to ascertain whether high grain quality was restricted to the locality in which Pusa is situated, how far the grain qualities were influenced by environmental factors, and how the milling and baking characters were affected by irrigation and by growth on the black soils. For this purpose, environmental investigations were carried out for some years at a number of stations situated in the wheat-growing areas. Briefly stated, it was found that in all the wheat-growing regions of India, including the canal-irrigated tracts of the Punjab and the black soils of the Peninsula, the essential milling and baking qualities of the Pusa wheats were maintained. In the case of Pusa 12, grown at thirteen stations in 1912, the milling and baking results obtained with the samples from the Indus valley and the black soil areas were actually better than those given by the Pusa sample and by those from other stations on the Gangetic alluvium. Canal irrigation was found to have no harmful effect on the grain quality, the best loaves being given by the wheat from Lyallpur. While environment had little effect on such characters as flour strength and the stability of the loaves, nevertheless the appearance of the samples, as judged by eye, was often adversely affected by the conditions under which they were grown. Thus in the set of samples of Pusa 12 grown in 1912, the Lyallpur lot was placed at the bottom of the list as judged by appearance and at the head of the list in the milling and baking tests. In his report Dr. Humphries stated: 'I should like to put side by side the two sequences of merit relating to these Pusa 12 wheats, one based on appearance in which all points of quality are taken into consideration, the other based on baking trials in which the only points of quality considered are stability and strength.' The two lists are as follows:

| *Appearance.* | *Strength and stability.* |
|---|---|
| Aligarh | Lyallpur |
| Meerut | Mirpurkhas |
| Cawnpore | Gurdaspur |
| Dumraon | Tharsa |
| Partabgarh | Cawnpore |
| Mirpurkhas | Pusa |
| Pusa | Aligarh |
| Gurdaspur | Meerut |
| Orai | Raipur |
| Tharsa | Bankipur |
| Raipur | Partabgarh |
| Bankipur | Orai |
| Lyallpur | Dumraon |

Having ascertained and repeatedly confirmed the fact that strong wheats with good milling qualities have been found to retain these characters under canal irrigation on the alluvium and also on the black soils of Peninsular India, the ground was cleared for further work. It then became possible to view the problem of the improvement of Indian wheat from a new standpoint. If this country can produce wheats of the same class as the best of those exported from North America, why should not a systematic effort be made to place India on a similar plane to Canada in the wheat markets of the world? It is evident to all who have even a superficial acquaintance with agricultural India and the Indian wheat trade, that such an enterprise would remain an idle dream unless the willing co-operation of the cultivator and of the Indian consumer could be enlisted. Any improvement in the grain itself, to be of importance, must satisfy both the Indian consumer and the Home miller. The Indian consumer as a rule exhibits a marked preference for a strong wheat. It was fortunate, therefore, that the class of wheat most liked by the people for food is that which is worth the most money on the Home markets. As soon as the new wheats were tried by the people they became popular for food purposes, and a substantial premium in the local markets was soon obtained. In the case of Pusa 12, this may be taken at six annas (6*d.*) a maund (82 lb.). In the case of Pusa 4 in Bundelkhand, a premium of as much as ten annas a maund is not uncommon. As regards the utilization of the grain there is therefore no clash of interests. India and Europe regard wheat from the same point of view. The Indian consumer is prepared to pay for quality. Export buyers, who wish to secure these wheats for shipment, will have to meet local competition.

The last step in the preliminary investigations was to determine whether yielding power, grain quality, rust-resistance, and strong straw could be combined in the same variety, when grown under cultivators' conditions, in such a manner that the systematic replacement of the country crop could be undertaken with success. No matter how high the grain quality and no matter the premium obtained for the seed, any new variety brought to the notice of the cultivator *must yield well under his conditions* and must be easily grown. Further, if grades of improved produce are to be established in India, the country crop over large areas must be replaced by one variety, so that the type can be kept reasonably pure and admixture rendered difficult. For this purpose, the new wheat must stand out in the field from the country varieties so that the extent of replacement can be readily observed and the degree of admixture of the crop easily detected. The working conditions, therefore, of this aspect of the question were the provision of varieties with good grain qualities which would yield well when grown by the people and which would stand out clearly from the country crop. Two wheats, Pusa 12 and Pusa 4, fulfilled these conditions and emerged from the preliminary trials as being the most useful types of the first set of Pusa wheats. The yield of both wheats was satisfactory when grown by the people. Pusa 12 is tall and beardless with shining red chaff, bright strong straw, and large white seed. Pusa 4 is a strong-strawed, beardless variety with bluish-green foliage, white felted chaff, and large translucent white grain Like Pusa 12 it differs sharply from the country wheats and is easily distinguished in the field.

Up to the present, the progress made in distributing improved wheats to the cultivators has been considerable. In the United Provinces, the area under Pusa 12 and Pusa 4 has reached over half a million acres and is rapidly expanding. Pusa 12 is being systematically distributed on the alluvial soils all over the Province, the largest area being found in Oudh and in the middle Doab in the Districts of Etawah, Cawnpore, Mainpuri, Fatehpur, and Farukkhabad. The distribution of Pusa 4 is largely confined to Bundelkhand—to the areas commanded by the Ken, Dhasan, and Betwa canals in the Districts of Banda, Hamirpur, and Jalaun—where an early variety is essential. In the North-West Frontier Province, the area under Pusa 4 is over 200,000 acres. In the Canal Colonies of the Punjab, the area under Punjab 11, a pure line selected by the Botanical Section, Pusa, from the local mixtures, now exceeds 750,000 acres. Outside North-West India, where the area under improved wheats is well into the second million acres, the distribution is proceeding and a number of schemes are being developed. Distribution of these wheats is in progress in Sind, in Bihar and Orissa, Bengal, Burma, the Central Provinces and Central India, Madras, Bombay, Gwalior, and the Rajputana States. The increased profit to the growers is at least fifteen rupees an acre, so that the annual dividend on this item of the work of the Agricultural Department is well over a million sterling a year and is rapidly increasing.

In some parts of the plains, where the wheat crop is liable to damage by birds and wild pigs, an improved bearded variety is desirable. A series of very promising bearded wheats, with strong straw and grain quality considerably better than that of Pusa 12, have been obtained by crossing Pusa 6 (high grain quality and rust-resistance) with Punjab 9 (stiff straw and general vigour). These new wheats have emerged successfully from the experiment station phase and are now being tried in the United Provinces under cultivators' conditions. One of these new types, Pusa 54, has been adopted for distribution in the Central Circle of the Province in those areas where a bearded wheat is essential.

On the black soils of Peninsular India, the replacement of the country crop by improved kinds has not been so successful as it has on the alluvium. In the first place, the soil and moisture conditions are such that only low yields can possibly be obtained under dry cultivation. In the second place, the soils are not so uniform as in the plains and the error of experiment in variety trials is considerable, a fact which makes it difficult to determine which variety is the best for general distribution. It is probable that a greater degree of success will attend efforts to improve the variety on the dry areas if more attention is paid to the type of ear and to the size of grain. As little or no tillering is possible, each plant cannot possibly form more than a limited number of grains, so it is important that these should be large and well developed. The best results are likely to be obtained by finding the most efficient types of macaroni wheats for certain of these areas.

Thus the problem of wheat improvement on the black soils is a twofold one—the discovery of more efficient macaroni wheats for many of the dry tracts and the development of bread wheats on the deep soil areas and under irrigation. On the deep moisture-retaining soils of the Narbada valley, the soft white wheats known as *pissi* have been studied by Evans, who has separated higher yielding types from the local mixtures. The seed of the best of these has been distributed to the cultivators and now covers large areas. In the eastern

Districts of the Central Provinces, the distribution of Pusa 100, a hybrid obtained by crossing Muzaffarnagar with a local Bihar wheat, is extending rapidly. In the dry tracts, however, the benefits to be derived from a change of variety must always, from the nature of the soil, be small. Irrigation united with intensive cultivation, however, alters the problem altogether, particularly on permeable soils. Growth is then stimulated in a perfectly marvellous way, and bread wheats develop the secondary root-system without difficulty. On some of the old poppy lands on the Malwa plateau of Central India, Pusa 4 yields from twenty-five to thirty-five maunds to the acre. Similar results are possible in other areas of the Peninsula provided sufficiently permeable soils are selected. The local macaroni wheats, however, are not very suitable for intensive cultivation, as they do not tiller sufficiently to render really high yields possible.

The immediate problem in wheat-production on the alluvium is the development of the means by which the present extensive system of agriculture can be intensified. For intensive cultivation the indigenous varieties are quite unsuitable. The new varieties, however, respond successfully to better soil conditions. Besides a suitable variety, two other things are required, namely, water and organic matter. The increase in production brought about by this means is extraordinary. The average yield of Pusa 12, gram, and sugar-cane obtained under intensive cultivation at Shahjahanpur for the seven-year period 1915–22 are given in maunds per acre in the next table :

*Average yields at Shahjahanpur under intensive cultivation.*

| *Crop.* | *Shahjahanpur.* | *Average yield obtained by cultivators.* |
|---|---|---|
| Wheat . . . . | 30·3 | 15·2 |
| Gram . . . . | 24·1 | 11·6 |
| Sugar-cane . . . | 841·0 | 345·4 |

These yields were obtained by the addition of organic manure, containing approximately 100 lb. of nitrogen per acre, to the sugar-cane crop once in the four-year rotation. The choice of the variety is important in the intensive cultivation of wheat. As is well known, the richer the land the greater the development of foliage and the weaker the straw. Enriching the land soon entails the lodging of the leafy indigenous types during the ripening period and a low yield of thin, poorly matured grain. The difficulty can be got over by selecting strong-strawed varieties with sparse foliage like Pusa 4 for really intensive cultivation. In such cases, the increased development of foliage on rich soil is not great enough to affect the standing power, and yields of over forty maunds to the acre have already been obtained both on the alluvium in Bihar and also on the black soils of Central India.

As the intensive cultivation of wheat spreads in India, a further improvement in production will become possible, namely, the saving of irrigation water. At Quetta in 1919, an acre plot in good condition gave on the preliminary irrigation before sowing, supplemented by 6·77 inches of winter rain, 2,886 lb. of grain and 4,715 lb. of straw. Similarly treated land but in much poorer condition yielded 1,625 lb. of grain and 2,745 lb. of straw. At Shaḥjahanpur in the same year, an area of 3·4 acres after sugar-cane in trenches gave the record yield of 36·5 maunds per acre on one irrigation. Irrigation water produces its maximum result if the moisture is made to cover the largest possible internal soil surface. The farther we can stretch the water, the more efficient it becomes. This is most easily achieved by increasing the supply of organic matter and by the preparation of a fine moist tilth after a preliminary irrigation. The crop must be carried as far as possible on this preliminary watering, after which only one or two further irrigations will be necessary. By these means, very large crops are possible with comparatively little water. Under present conditions, when canal water is assessed by the area irrigated and not by the amount used, there is no inducement for the cultivator to economize. Everything is now pointing to a change in irrigation policy. The adoption of volumetric methods of sale cannot long be delayed. When this takes place, the cultivator will rapidly discover for himself that better cultivation means less water, and he will be quick to reduce this item of his expenditure.

## BIBLIOGRAPHY

Burt, B. C.—The suitability of Pusa 12 wheat for local consumption in the United Provinces, *Agr. Jour. of India*, x, 1915, p. 370.

Burt, B. C., Howard, A., and Howard, G. L. C.—Pusa 12 and Pusa 4 in the Central Circle of the United Provinces, *Bull. 122, Agr. Research Institute, Pusa*, 1921.

Clarke, G., Banerjee, S. C., Naib Husain, M., and Qayum, A.—Nitrate fluctuation in the Gangetic alluvium and some aspects of the nitrogen problem in India, *Agr. Jour. of India*, xvii, 1922, p. 463.

Clouston, D.—Manurial experiments with wheat at the Nagpur Experimental Farm, *Agr. Jour. of India*, i, 1906, p. 205.

Evans, G.—The improvement of the seed supply in the Central Provinces, *Agr. Jour. of India*, xii, 1917, p. 58.

Hailey, H. R. C.—Private farms in Oudh, *Agr. Jour. of India*, xii, 1917, p. 519.

Howard, A.—The influence of the weather on the yield of wheat, *Agr. Jour. of India*, xi, 1916, p. 351.

——————Pusa 12, *Agr. Jour. of India*, x, 1915, p. 1.

Howard, G. L. C.—The wheats of Baluchistan, Khorasan, and the Kurram valley, *Mem. of the Dept. of Agr. in India* (*Botanical Series*), viii, 1916, p. 1.

Howard, A., and Howard, G. L. C.—The varietal characters of Indian wheats, *Mem. of the Dept. of Agr. in India* (*Botanical Series*), ii, 1908.

——————Wheat in India, Calcutta, 1909.

——————The milling and baking qualities of Indian wheats, *Bulls. 14, 17, and 22, Agr. Research Institute, Pusa*, 1908–11.

——————The improvement of Indian wheat, *Agr. Jour. of India*, viii, 1913, p. 1.

——————The agricultural development of North-West India, *Agr. Jour. of India* (*Indian Science Congress Number*), xii, 1917, p. 67.

——————The saving of irrigation water in wheat growing, *Bull. 118, Agr. Research Institute, Pusa*, 1921.

Howard, A., Howard, G. L. C., and Abdur Rahman Khan.—The economic significance of natural cross-fertilization in India, *Mem. of the Dept. of Agr. in India* (*Botanical Series*), iii, 1910, p. 281.

——————The wheats of Bihar and Orissa, *Mem. of the Dept. of Agr. in India* (*Botanical Series*), xii, 1922, p. 1.

Howard, A., Leake, H. M., and Howard, G. L. C.—The influence of the environment on the milling and baking qualities of wheat in India,

*Mem. of the Dept. of Agr. in India (Botanical Series)*, iii, 1910, p. 191; v, 1913, p. 49; and vi, 1914, p. 233.

Main, T. F.—Irrigation in Sind, *Agr. Jour. of India*, xiii, 1918, p. 653.

Moreland, W. H.—The Oudh seed dépôts, *Agr. Jour. of India*, ii, 1907, p. 217.

Wilson, J.—Memorandum on Indian wheat for the British market, *Bull. 20, Agr. Research Institute, Pusa*, 1910.

# XII

## RICE

ALTHOUGH rice, one of the chief cereals of the torrid zone, is cultivated in all tropical countries, nevertheless 90 per cent. of the world's crop is grown in Eastern and South-Eastern Asia—in India, Japan, China, Java, Indo-China, Siam, and Chosen (Korea). Excluding China, for which statistics are not available, India produces from three-fifths to two-thirds of the rice of the world. During the period 1900 to 1921, the annual production of this country was approximately 35,000,000 tons of cleaned rice. The total area in 1922 was estimated at 81,535,000 acres and the average yield 910 lb. per acre. Bengal stands first among the rice-producing Provinces, the area sown being approximately a quarter of that of the whole Indian Empire. Bihar and Orissa is not far behind, while the area in both Madras and Burma is about one-eighth of the total production. The rice crop of Bengal is therefore about equal to that of Madras and Burma combined. In the rest of the Provinces, although the area sown is considerable, rice is a comparatively unimportant crop.

Burma is by far the chief exporting Province, and shipped in 1922 no less than 1,745,000 tons out of the total Indian export of 1,999,500 tons. The growth of the shipments of rice from Rangoon has been largely a consequence of the discovery of the most efficient method of removing the husk of the paddy and at the same time providing cheap fuel (in the form of the rice husk) for the mills. Removal by hand was out of the question on account of the dearness of labour and the disinclination of the Burmese cultivator to undertake any work which he can avoid. Hence the introduction of the power-mill and the growth of the rice-milling industry of Rangoon. This is now so efficient that it no longer pays to transport rice in husk to be milled at places remote from the areas of production. As in the case of the jute crop, the energy and enterprise of the local merchants have led to important developments in agriculture and have provided the Burmese cultivator with an easily grown and easily marketed money crop. The removal of the husk by means of small power mills is now spreading in Bengal and in other parts of India, and may be taken as a definite proof of the growing prosperity of the people. The spread of rice-milling in India is bound, as in Burma, to influence the varieties of paddy grown. Types with uniform, bold, rounded grains, which pass through the mill with little breakage, give the highest yield of finished rice.

The requirements of the rice plant are an abundant supply of fresh water (rain or irrigation) during the growing season, a high temperature, and a level tract of rather heavy soil, underlain by a subsoil that is largely impervious to water. The crop is partly submerged for about three months of its existence, so that the volume of water required for maturation is very considerable. Rice is an aquatic plant, and therefore differs fundamentally from most of the other cultivated crops of India. Similar swamp cultivation is hardly represented in Western Europe, the only crops at all like rice being osier willows and water-cress beds, both of which are not at all common and have received little or no attention on the part of agricultural chemists. When, therefore, the investigators of the Agricultural Department in India came in contact with rice, they found a set of soil conditions quite different from anything they had previously experienced. Beyond the traditional knowledge of the people, which is entirely empirical, none of the principles underlying the cultivation of a swamp plant had been ascertained. Even in 1921, when the last edition of Russell's *Soil Conditions and Plant Growth* was published, less than two out of the 400 pages are devoted to the agricultural chemistry of the rice field. Although our knowledge of the fundamental principles underlying the production of rice is still fragmentary, nevertheless a great deal has been done in India in unravelling and defining the problems involved. In addition, a large amount of successful work has been accomplished in increasing the production of rice. The chief problems which remain to be solved, as well as the most striking results which have been obtained, will now be dealt with, and it will be convenient to consider them under two main heads—the soil and the plant.

In any scientific consideration of the soils which produce the rice crop in India, practical experience at first seems to contradict one of the great principles of the agricultural science of the Occident, namely, the dependence of cereals on nitrogenous manures. Large rice crops are produced in many parts of India on the same land year after year, without the addition of any manure whatever. The rice fields of the country export paddy in large quantities to the centres of population or abroad, and there is no corresponding import of combined nitrogen. Taking Burma as an example of an area exporting rice beyond seas, during the last twenty years about 25,000,000 tons of paddy have been exported from a tract roughly 10,000,000 acres in area. As unhusked rice contains about 1·2 per cent. of nitrogen, the amount of this element shipped overseas during the last twenty years or destroyed by the burning of the husk is in the neighbourhood of 300,000 tons. As this constant drain of nitrogen is not made up by the import of manure, we should expect to find a gradual loss of fertility. Nevertheless this does not take place either in Burma or in Bengal, where rice has been grown on the same land year after year for centuries. Clearly the soil must obtain fresh supplies of nitrogen from somewhere, otherwise the crop would cease to grow. The only likely source is fixation from the atmosphere, and the question arises how can this take place in a Bengal rice field, for example. Some fixation is possible in the cold weather after the rice is cut, but how does the crop obtain the large quantity of nitrogen it is known to take up between transplanting and seed formation? Jatindra Nath Sen has shown that the rice plant during this period increases in total nitrogen content more than ten times. It seems probable, therefore, that actual fixation must take place in the rice fields themselves while the land is under water. The most probable seat of this fixation is in the submerged algal film on the surface of the mud. We know from the work of Brizi and his successors that one important rôle of the algal film is to supply oxygen in solution to the roots of the rice plant. Harrison and Subramania Aiyer have shown that nitrogen gas is one of the products of fermentation in rice soils. During the period when the rice plant is gaining in nitrogen the evolution of this gas from the cropped soils is much less than during the flowering period, by which time the plant has ceased to assimilate this nutrient. This point is most suggestive, and requires further study. It may materially help in the solution of the problem of the nitrogen supply of rice. Now the algal film contains bacteria in addition to green plants. It is quite possible therefore that the algae and bacteria in the film, working conjointly, are capable of using the energy of sunlight to fix some of the nitrogen evolved in fermentation in a form suitable for the rice plant. The sources of all the nitrogen taken up by the rice crop in India, the forms in which it is absorbed by the plant at different stages of its growth, and the complete nitrogen cycle of the rice fields are matters in which we are almost completely in the dark. The only definite facts so far ascertained are: (1) the large amount of nitrogen taken up by the rice crop between transplanting and seed-formation, and (2) the preference of the rice plant for ammonia rather than nitrate as a source of its nitrogen. After a supply of combined nitrogen, the next most important requirement of the roots of the rice crop is dissolved oxygen, and here our knowledge, thanks to the researches of Harrison and Aiyer, is in a much more satisfactory condition than is the case with the nitrogen supply. These investigators have worked out in detail the manner in which the algal film is able to bring about a continuous aeration of the roots of the rice plant and of the upper layers of the submerged soil. Apart from the value of these researches in explaining how the aeration of the rice roots takes place, they have helped to bring into prominence the fact that in the production of rice the plant and the soil are only two of the factors involved. Besides these the algae and bacteria in the water above the mud are just as important as the plant or the soil. The requirements of the algal film as regards minerals and light, the part it plays in the growth of the crop, in such operations as green-manuring and the various methods by which its activities can be increased, are matters which may easily prove as important in rice production as the selection of a higher yielding variety of paddy. Rice cultivation is therefore an exceedingly complex thing, of which the principles have only in part been investigated. It is hoped that the Calcutta School of Chemistry will be able to relieve the

Agricultural Department of a portion of its burden and render available a detailed scientific account of the various processes taking place in the rice soils of Bengal. When this is done, a new and most interesting chapter will have been added to agricultural science.

Passing from the activities of the plant and of the various living organisms met with in the rice fields, the texture of the soil is of great importance in the cultivation of rice. This factor affects the plant in two different ways. The permeability of the soil must be destroyed to a large extent by the puddling process, otherwise the rice fields will not retain water. On the other hand, the mud in the surface layers of soil must be sufficiently permeable to admit of the passage of air and dissolved food materials to the roots. Unless the exact stage of very slow drainage is reached, the field will either lose too much water by percolation or efficient root action will become impossible through inadequate aeration. The destruction of the permeability must be carried to a certain point, and drainage must not be altogether prevented. While it is not difficult to obtain a suitable texture at the time of transplanting, it may not prove easy to maintain this condition till the crop ripens. It is more than probable that the little permeability there is to begin with may tend to disappear altogether, after which efficient root activity will become more difficult. Some easy method of maintaining or increasing the necessary slow drainage may be needed. Now soil texture is readily influenced by the addition of green-manure and also by small quantities of salts, such as sulphates, so that it becomes important to consider all the empirical results obtained in India by means of green-manure and of substances like sodium sulphate, magnesium sulphate, superphosphate, and bone meal (either alone or in combination with organic matter) from the point of view of their possible effect on the texture and permeability of the soil. The optimum texture of the soil of the rice fields at all stages of growth—both from the point of view of the plant and of the algal film—and the cheapest methods of obtaining it are subjects which need investigation as urgently as the sources of the nitrogen used by the rice crop.

Of equal importance to the agricultural chemistry of the

rice fields themselves (both before[1] and during the life of the crop) is the detailed study of the rice nursery. As is well known, transplanted rice usually gives a better yield than the same variety broadcasted, but the full explanation of this curious result has still to be determined. The operation of transplanting is attended by a number of disadvantages as far as the plant is concerned. It involves the complete loss of the original root-system, the formation of a new one largely at the expense of the seedling, and, what is still more important, a great loss of valuable time. Nevertheless, in the end, the process pays. Why is this? One explanation which has been advanced is the increased efficiency of the new root-system, but against this is the well-marked power of aquatic plants to adjust their root-development to the working conditions. If it pays transplanted rice to develop roots near the surface of the mud, there is no reason why these roots should not be formed in the case of broadcasted paddy. One possible explanation of the benefits of transplanting is suggested when the results obtained by the *rab* process in Western India and the work of Jatindra Nath Sen, in his study of the assimilation of nutrients by the rice plant, are considered together. On the stiff soils of Western India, the cultivators bring the intractable soil of the rice nurseries into a suitable condition for the seedlings by burning cow-dung and the leaves and branches of forest trees on the surface of the ground just before sowing. This operation is known as *rab*. The researches of Mann, Joshi, and Kanitkar have shown that the beneficial result of *rab* is partly due to the improved texture of the soil after it has been heated, and partly to the effect of the ashes. Safflower cake has a similar effect to that of heat on the soil colloids, and has proved an efficient *rab* substitute. The effect of this troublesome and expensive process on the rice seedling is twofold. Growth in the nursery is improved and the resulting crop also benefits. In other parts of India, the cultivators pay great attention to

[1] In some parts of India, such as North and South Bihar, it is an advantage to cultivate the rice fields after harvest; in others, such as Mysore, such cultivation does not pay. The efficiency of this cultivation evidently varies with the soil type, but the reasons for success and failure in the two cases have not yet been determined.

the rice seed-beds, and apply comparatively large quantities of cow-dung to them. In the nursery stage the rice plant needs a good tilth, and the soil must also be in excellent condition. It therefore behaves like a dry crop, when this is grown intensively. In addition, the seedlings must be a suitable size before they are transplanted. Now Jatindra Nath Sen has shown that the rice seedling is rich in nitrogen, much richer in fact than at any further stage of its existence. This fact throws a flood of light on the real meaning of the *rab* process and on the care taken by the cultivators generally in the preparation of the rice nursery. Clearly before the aquatic phase of the rice crop begins, the seedling must be heavily charged with nitrogen. It is possible that the young plant cannot secure this in sufficient amount when the crop is broadcasted, and can only obtain this essential food material in quantity in the nursery stage. If this is so, the advantages of transplanting would at once be explained. The nursery phase of the rice crop should receive the fullest investigation, and a comparison should be made between the composition of the seedlings sown broadcast and those raised in seed-beds. The Madras Agricultural Department has already demonstrated the saving of seed that results from reducing the seedlings from a bunch to one in transplanting, but little work has been done to determine the most efficient size of the seedling, or whether any improvement is possible in the management of rice nurseries. The advantages of transplanting over broadcasting have been clearly demonstrated in Chhattisgarh, where, through the efforts of Clouston and his staff, this practice is now widespread.

On the general question of manures for rice, the numerous experiments which have been carried out in India point to the great value of organic manure, including green-manure, in increasing production. These results are in accordance with experience, as the benefit of incorporating the weeds into the mud when the rice fields are puddled is well known to the cultivators. The addition of green-manure or cow-dung is nothing more than an improvement on an age-long custom. Besides demonstrating the beneficial effects of green-manure for transplanted paddy, the Agricultural Department in Orissa is introducing this process in the case of the broadcasted crop. The seeds of the rice and of the green-manure crop are sown together. At a later stage, the land is ploughed and cross-ploughed. Many of the rice seedlings and most of the green-manure plants are then destroyed, but the paddy which survives rapidly improves, as a result of the addition of organic matter at this stage of its growth. In Bengal, large amounts of green-manure, in the shape of the water hyacinth, are available on the spot, but much further work is needed to discover the best methods of utilizing this natural green-manure in the improvement of rice-production.

The botanical aspects of rice-cultivation have received considerable attention on the part of the Agricultural Department, and already very valuable results have been obtained. In the absence of an all-India rice experiment station for the study of the fundamental principles underlying the subject, the botanical work on this crop has fallen to the Provincial Departments. Work is now in progress in Bengal, Madras, Burma, the Central Provinces, Bombay, and in Bihar and Orissa. At practically all the rice stations, the Economic Botanists are actively engaged in the isolation of high-yielding unit species from the mixtures grown, in the testing of promising types, and in growing seed for distribution to the cultivators. This method of improvement is possible, as the amount of natural cross-fertilization in rice is small. In most of the Provinces, notably in Bengal, Burma, the Central Provinces, and Bihar and Orissa, results of practical value have been obtained, and the seed distributed to the cultivators gives a distinctly better yield than the local crop. The best known of the improved varieties is Indrasail (isolated by Hector at Dacca), which covers thousands of acres in Bengal and has also spread to Assam and to Bihar. One point in connexion with the selection work on rice has not been emphasized in the annual reports, namely, the great importance of checking the district work by means of some easily recognized morphological character in the new types. Unless the new rices stand out clearly from the ordinary crop, it is very difficult, if not impossible, to keep the seed pure on the seed-farms and to obtain reliable estimates of the area under the new types. Although in the large deltaic regions of India and Burma, where the soil and other conditions are very uniform, it has been possible to isolate pure lines which yield better than the mixed crop, it by no means follows that this method will always give the best results. The researches of Koch in Java point to the failure of the unit species the moment the soil and moisture conditions vary. Suitable mixtures there do better than a single type The Java experience should, therefore, not be forgotten in those parts of India where soil and water-supply are apt to be less uniform than in tracts like Bengal, Burma, and the deltas of Madras. In such areas, a fortunate combination of forms is likely to do better than a single type. Besides the work on form separation, progress has been made at Dacca and Coimbatore in building up a solid foundation of knowledge of the inheritance of characters in rice, the practical value of which will become evident in the later stages of rice improvement.

The work in progress on the separation of better rices from the country crop by no means exhausts the work before the botanist. Except in the Central Provinces, where both the cultivated and wild rices have been classified, no detailed botanical survey of the Indian rice crop has yet been made, and little or no attention has been paid to the various species and varieties related to the ordinary rices of cultivation. Such a botanical survey is an immense undertaking, but it is essential before the further stages in the improvement in the crop can be undertaken. After the best indigenous types have been separated and the country crop has been replaced by more efficient kinds, the possibilities of hybridization will have to be explored. With the more intensive rice cultivation essential to support the rapidly growing population and to meet the constantly improving standard of living, forms with much stronger straw and better root-development will be needed. To produce these, the botanist will require the largest possible armoury of varieties, and it will probably happen that wild types, quite useless in themselves, will contribute useful characters to the rices of the future. The creation of these will involve not only a profound knowledge of existing forms and of their root-development, but also of the way the various characters are inherited. At present, nothing is known of the most efficient type of root-system for early, late, and deep-water rices, and till this has been worked out, the plant breeders are to a certain extent working in the dark. Besides a knowledge of root-development, the rice improvers of the future will have to combine their knowledge of the crop with all that is known of the physiology of the plant, with the relations of the crop to the soil and to cultural methods, so as to evolve improved types not only suitable for large areas, but able also to respond effectively to better soil conditions. In considering the diseases of rice, such a wide outlook will be equally important. In the complex system of mutually interacting factors which a rice field presents, it is obvious that something more than a special knowledge of some particular parasite will be needed in the study of disease. When for example in Bengal, one of the greatest centres of rice-production in the world, disease appears, the whole of the complex mechanism of production should be examined in detail as well as the life-history of the supposed parasite. Such a survey is desirable at the present time in the case of the *ufra* disease of rice, which occurs towards the mouths of the Brahmaputra in Eastern Bengal. Here broadcasted deep-water paddy is attacked by an eel-worm (*Tylenchus angustus* Butler), but the transplanted crop alongside escapes. A little consideration will show that if this eel-worm is a real enemy of rice, it would have spread rapidly all over Bengal and affected the whole of the rice crop in this tract. The fact that only the deep-water broadcasted paddy is attacked and that the neighbouring transplanted rice escapes suggests the desirability of a careful study of all the factors involved in the growth of broadcasted rice in this tract. Something is almost certain to be wrong with the root-system, as a very similar eel-worm attacks wheat when this crop is grown in stiff, poorly cultivated, and over-irrigated soils on the Western Frontier. The neighbouring fields which have been properly cultivated escape. It is possible either that the rice seedlings of the broadcasted crop are unable to obtain sufficient nitrogen and other food materials in the early stages of growth or that the efficiency of the algal film is being interfered with by some adverse factor such as want of light. The results described suggest that the crop is suffering from some form of mal-nutrition and that the attack of the eel-worm is a consequence of some kind of starvation. It is difficult to regard the worm as the primary cause of the *ufra* disease.

## BIBLIOGRAPHY

Basu, S. K.—Green-manuring of broadcasted paddy in Orissa, *Agr. Jour. of India*, xvi, 1921, p. 689.

Bristol, B. M., and Page, H. J.—A critical inquiry into the alleged fixation of nitrogen by green algae, *Ann. of Applied Biology*, x, 1923, p. 378.

Butler, E. J.—Diseases of Rice, *Bull. 34, Agr. Research Institute, Pusa*, 1913.

————The cultivation of rice in Spain and the recent International Rice Congress at Valencia, *Agr. Jour. of India*, ix, 1914, p. 326.

————The rice worm (*Tylenchus angustus*) and its control, *Mem. of the Dept. of Agr. in India* (*Botanical Series*), x, 1919, p. 1.

Clouston, D.—The transplanting of rice in Chattisgarh, *Agr. Jour. of India*, iii, 1908, p. 338.

Coleman, L. C., and Ramchandra Rao, D. G.—Experiments on paddy cultivation during the years 1909–11, Bangalore, 1912.

Fruwirth, C.—Handbuch der landw. Pflanzenzüchtung, Bd. v, Berlin, 1923.

Graham, R. J. D.—Preliminary note on the classification of rice in the Central Provinces, *Mem. of the Dept. of Agr. in India* (*Botanical Series*), vi, 1913, p. 209.

Harrison, W. H., and Subramania Aiyer, P. A.—The gases of swamp rice soils, *Mem. of the Dept. of Agr. in India* (*Chemical Series*), iii, 1913, p. 65; iv, 1914, p. 1; iv, 1916, p. 135; v, 1916, p. 1; and v, 1920, p. 173.

Hector, G. P.—Notes on pollination and cross-fertilization in the common rice plant, *Oryza sativa* L., *Mem. of the Dept. of Agr. in India* (*Botanical Series*), vi, 1913, p. 1.

————Observations on the inheritance of anthocyan pigment in paddy varieties, *Mem. of the Dept. of Agr. in India* (*Botanical Series*), viii, 1916, p. 89; and xi, 1922, p. 153.

Joshi, P. G.—Rice growing in the Koukan without transplanting, *Agr. Jour. of India*, xix, 1924, p. 160.

Kikkawa, S.—On the classification of cultivated rice, *Jour. of the College of Agr., Tokyo*, iii, 1910, pp. 11 et seq.

Kossowitsch, P.—Untersuchungen uber die Frage, ob die Algen freien Stickstoff fixiren, *Bot. Zeit.*, v, 1894, s. 98.

McKerral, A.—Some problems in rice improvement in Burma, *Agr. Jour. of India*, viii, 1913, p. 317.

Mann, H. H., Joshi, N. V., and Kanitkar, N. V.—The *rab* system of rice cultivation in Western India, *Mem. of the Dept. of Agr. in India (Chemical Series)*, ii, 1912, p. 141.

Parnell, F. R.—The inheritance of characters in rice, *Mem. of the Dept. of Agr. in India (Botanical Series)*, ix, 1919, p. 75, and xi, 1922, p. 185.

————————Note on the detection of segregation by examination of the pollen of rice, *Jour. of Genetics*, xi, 1921, p. 209.

Paton, Noel.—Burma Rice, Calcutta, 1912.

Roy, S. C.—A preliminary classification of the wild rices of the Central Provinces and Berar, *Agr. Jour. of India*, xvi, 1921, p. 365.

Sen, J. N.—A study in the assimilation of nutrients by the rice plant, *Bull. 65, Agr. Research Institute, Pusa*, 1916.

Sherrard, G. C.—A summary of the experiments on rice in Bihar and Orissa from 1912 to 1919, *Bull. 96, Agr. Research Institute, Pusa*, 1920.

Taylor, C. S., and Manmathanath Ghose.—Experiments in the green-manuring of rice, *Agr. Jour. of India*, xviii, 1923, p. 104.

Warth, F. J., and Maung Po Shin.—The phosphate requirements of some Lower Burma paddy soils, *Mem. of the Dept. of Agr. in India (Chemical Series)*, v, 1919, p. 131.

Watt, G.—The Commercial Products of India, London, 1908.

# XIII

## SUGAR-CANE

Although sugar-cane is cultivated in all parts of India and the total area under this crop is nearly two and three-quarter million acres, nevertheless production and consumption do not balance. India is an importing country as far as sugar is concerned, and every year has to spend large sums in the form of exports to make up for the shortage. One reason for this unsatisfactory state of affairs is the extraordinary low standard of production, compared with that of the other cane-growing regions of the world. The average yield of sugar in India, during the five years ending 1918–19, was about 1 ton to the acre compared with 2 tons in Cuba, over 4 tons in Java, and more than 4½ tons in Hawaii. During the four pre-war years, 1910 to 1914, the imports of sugar into India averaged 723,915 tons, valued at 12·71 crores of rupees. Of this, by far the greater part was contributed by Java, the rest chiefly by Mauritius and Austria-Hungary. With regard to sugar, therefore, a serious economic drain is in active progress, and no effort should be spared in exploring the best means of making India self-supporting in this respect.

The distribution of the sugar-cane crop in India presents several interesting features. Although cane is essentially a plant of the tropics, nevertheless 75 per cent. of the area is found in the Indo-Gangetic plain, to the north of the tropic of Cancer. The area under cane in tropical India is surprisingly small, and moreover shows little or no signs of expansion. Fifty per cent. of the Indian crop is found in the United Provinces, 15 per cent. in the neighbouring Districts of the Punjab, and 10 per cent. in North Bihar. The great sugar tract of India occurs on the alluvium of the great plains in a wide belt parallel to and at no great distance from the Himalayas, stretching from Gurdaspur on the west to Darbhanga on the east. Temperature and moisture appear to be the chief factors which limit this crop in the sub-montane tracts of the Punjab. The eastward extension of the cane area in Bihar seems to be checked by two circumstances—poor soil-aeration and the rapid increase of the rain-inundated area in which rice is naturally a more suitable crop. The main factors responsible for the marked concentration of sugar-cane in the sub-montane areas appear to be soil moisture, humidity, and temperature. There are three tracts in the United Provinces where sugar-cane exceeds five per cent. of the net area cropped. The first of these comprises the Districts of Saharanpur, Muzaffarnagar, Bijnor, Meerut, and Bulandshahr, all situated on the great drainage lines of the Jumna and the Ganges. The second of these important sugar tracts comprises the Districts of Bareilly and Pilibhit in Rohilkhand. The third occurs towards the east and embraces the Districts of Gorakhpur, Fyzabad, Azamgarh, Jaunpur, and Ballia. From the general lie of the country, the distribution of the rivers, and the position of the subsoil water-level, it seems probable that the soil moisture conditions in these three regions favour the root development of the cane and so help to reduce the amount of irrigation water needed by the crop. It would be interesting to explore in detail the various factors responsible for the concentration of the cane crop in these areas, as the information might easily prove to be of value in practice. Another arresting fact connected with the distribution of cane in India is the small amount of attention devoted to this crop in the rice areas of Madras, although excellent crops of thick cane can be raised there without difficulty. The cultivator in these regions appears to be imbued with a strong instinctive preference for rice, although the profits obtainable from sugar are very large. In the deltaic tracts of the Peninsula, the extension of irrigation facilities has always developed the cultivation of rice rather than the production of sugar. So far, the scientific basis of this marked preference for paddy has not been investigated. It is probable that the ryots' preference for rice has a real scientific basis. Paddy lands are self-sufficient as far as nitrogen is concerned. Sugar-cane needs expensive nitrogenous manure. It is only on the new canals of the plateau, when water is provided by the State, that the cultivators of tropical India turn to the growth of sugar-cane.

The general facts of distribution and the present condition of the industry all point to the United Provinces as being the centre of the subject as far as the improvement of sugar-production in India is concerned. The area under cane is already considerable, and sugar is perhaps the most important money crop cultivated. The lines of future development have already been worked out by Clarke at the Sugar Experiment Station at Shahjahanpur, while a large and important irrigation system will shortly be in operation in Rohilkhand and Oudh which will remove one of the limiting factors which has hitherto hindered development, namely, a supply of cheap irrigation water. A supply of water at moderate rates will also help to mitigate another unfavourable factor in cane cultivation in this tract—the low atmospheric humidity which obtains during the early period of the crop from the harvest in March to the break of the rains in June. This factor not only checks the rate of growth but also increases the cost of production very considerably. The cultivator strives to avoid the harmful effects of low humidity during the hot months by close planting and by the cultivation of narrow-leaved types of cane, but in spite of this and the provision of irrigation, the crop makes little growth before the rains. Another unfavourable factor in sugar-production in the United Provinces is the shortness of the growth period. The cane is planted in February and the crop establishes itself during the hot season. Only during the rains and till November is growth rapid and really satisfactory. After November the cold weather sets in, growth is checked, the ripening process is completed, and the cane is crushed. Really high yields under such circumstances are clearly impossible, as cropping power must be sacrificed to early maturity. It is easy to understand, therefore, that, compared with such highly favoured localities as Java and Cuba, not only is the yield of cane per acre relatively low but the cost per ton is high.

The general line of advance appears to lie in increasing the general production of cane and of juice per unit area, at the same time reducing the cost of each ton of cane grown. This can be achieved either by substituting more efficient varieties for those now grown, or still better by the more intensive cultivation of improved varieties. What is required is a rapidly growing and rapidly maturing cane, with a root-system suited to the soil conditions and vegetative vigour and disease-resistance above the average. The ratio of juice to cane should be high, and the quality of the juice should allow of the easy manufacture of high-grade country sugar. Afterwards further advances can be made by simple improvements in crushing which are now possible by the use of small oil engines in place of oxen and buffaloes. The manner in which such advances can best be made is discussed in the recent *Report of the Sugar Committee*, a volume which should be carefully studied by all agricultural investigators in India and by advanced students of the subject. It indicates a clear grasp of the problems to be solved as well as a high degree of agricultural insight. Although all of its findings cannot be accepted without some qualification, and in one or two instances essential factors have been forgotten, this report serves as an admirable starting-point for the further work which is necessary if any real progress is to be made. One of the most illuminating sections of the report is the one dealing with the cultivation of the cane in Java. In that island, rice lands are leased by the sugar planters under regulations laid down by Government. Cane follows rice, and yields of over forty tons to the acre are common. Considering the character and exhaustive nature of the rotation and the absence of fallows, surprisingly little nitrogenous manure—about 80 lb. of nitrogen to the acre on the average—is applied to the soil for the cane crop. Nevertheless, there is no falling off in general fertility, the cost of cane working out at less than five annas a maund. In Java the greatest attention is paid to two things—the variety and the maintenance of the aeration of the soil. There is an efficient system of surface-drainage, the canes are grown in trenches and afterwards earthed up periodically, and, where irrigation is practised, the very minimum quantity of water is used. The cane roots thus obtain an abundant air-supply and, what is perhaps more important, there is ample aeration for the soil organisms engaged in nitrogen fixation. The Dutch planters are most careful to allow the soil to manure itself, and they only supplement the natural recuperative agencies at work. Besides an efficient agriculture, the manufacture of the cane and the sale of the produce are highly organized. The planters, their research officers, and the Government have all united to make the industry a success. Two other facts of interest to India are to be found in this account of the Java sugar industry. Firstly, the scientific investigators employed by the planters formerly included an entomologist and a mycologist, but these are no longer considered necessary, as it is held that proper methods of cultivation and the introduction of good varieties are the most important factors in the control and elimination of diseases. Secondly, two systems of cane cultivation exist in Java. Side by side with the efficient production of cane by the planters, there is an indigenous cultivation, the yield of which is not more than half that obtained by the sugar factories. The yield of raw sugar per acre obtained by the Javanese is about the same as that obtained in India, namely, about a ton to the acre. This fact is most significant, and suggests that, in addition to example, active propaganda will always be needed in developing agricultural production in the Orient.

The problems which await solution before the production of sugar in India can be increased are fortunately simpler than those which underlie progress in cotton cultivation. Some of these will now be considered.

Before any attempt is made to introduce new varieties to the notice of the cultivator, it is obviously essential that the investigator should know what kinds are already grown, the relation of these types to local soils and local conditions, and in what directions they need improvement. This is the work of a botanical survey of the existing crop, carried out from the agricultural standpoint. Fortunately, a beginning has been made by Barber and others, and large collections of indigenous canes are already in existence. One important omission in this survey, which however is by no means complete, is a study of the root-systems of the unit species and their relation to the soil conditions throughout the period of growth. A beginning, however, has recently been made in this direction by Venkatraman, but an enormous amount of work still remains to be done. A knowledge of the life-history of the root-system of the cane is essential before we can understand the general geographical distribution of the crop, the occurrence of types like *Saretha* in the Meerut Division and *Hemja* (*Burli*) in Gorakhpur and North Bihar. It would also be useful in helping to indicate the tracts in which new varieties are most likely to succeed. It is said that both the *Saretha* and *Hemja* groups of cane deteriorate when grown outside their respective areas. The cause of these changes might be better understood if we knew more about the details of root-development of these types. In Bihar, *Hemja* is planted in February and actually grows without irrigation during the hot months of April and May, when even deep-rooted crops like Java indigo often maintain themselves with difficulty. Perhaps this variety develops a specially deep root-system during the hot months and a more superficial one for the rains after the aeration of the subsoil falls off. Another direction of profitable inquiry is the relation between earthing-up the cane and root-development. Earthing-up increases the yield and improves the standing power. These results are probably due not only to the effect of this operation on the development of a constant succession of new roots from the base of the cane, but also to the continuous supplies of food materials rendered possible by the maintenance of efficient soil-aeration.

The search for more efficient varieties of cane became systematized after the 1911 meeting of the Board of Agriculture, when a definite recommendation was made that a cane-breeding station should be opened in Madras for the purpose of raising new seedlings for the plains. The general lines of work for the new station were laid down at this meeting and were shortly afterwards approved by the Government of India. Since 1912, a large amount of valuable work has been done at Coimbatore by Barber and Venkatraman. A botanical survey of the indigenous canes has been started, and a detailed study of the facts connected with the raising and testing of seedling canes has been completed. The choice of Coimbatore as a suitable centre for the production of seedlings has been amply justified. A number of new canes, particularly those in which *Saccharum spontaneum* L. was one of the parents, are proving successful in various parts of Northern India. In North Bihar, the distribution of the best of these new types is being actively developed from Pusa by the recently formed Sugar Bureau. In the Punjab and the United Provinces, the new seedlings are doing well on the Government farms and their distribution to the cultivators is beginning.

At Shahjahanpur, in addition to an exhaustive trial of the new Coimbatore seedlings, a search for better canes is being made in two other directions, namely (1) the isolation of pure lines found in the indigenous mixtures, and (2) the importation of varieties from overseas. Very promising results have been obtained, and some of the imported varieties are being rapidly taken up by the cultivators. There is one further possibility in the search for a better cane for the plains which still remains to be investigated, namely, the raising of the seedling canes in the United Provinces itself. At present, seedling canes for the whole of India are raised at Coimbatore. No effective preliminary selection of the seedlings suitable for Northern India is, however, possible in South India, so that a mass of material, of necessity mainly unsuitable, has to be sent to the plains for trial. As the indigenous canes sometimes flower freely in the United Provinces at the beginning of the hot weather, it should not prove impossible to make them produce pollen and also fertile seed. The air is too dry at flowering time for the anthers to complete their development, but the humidity could be raised by growing the experimental canes in a chick house or in a betel-leaf garden. Only a few cane stools are needed for the production of seedlings, so that the expense and trouble in working out the proper soil and atmospheric conditions for seed-formation in the United Provinces need not be considerable. If this proves successful, the activities of the Coimbatore station could be devoted to Southern and Peninsular India.

After the provision of more efficient canes, it is important to maintain the new varieties as long as possible. As the cane is vegetatively propagated, no loss of vigour through vicinism is possible. New cane varieties, like new kinds of potatoes, do, however, deteriorate, but so far we do not know how this takes place and why continuous vegetative propagation often leads to a gradual loss of vigour. This is one of the fundamental questions underlying agriculture which needs investigation, and which can only be undertaken at Experiment Stations maintained for long periods by the State. It is a question which certainly should be taken up in India without delay. So far, the only known method of prolonging the efficiency of the variety is a periodical supply of new sets brought from a colder locality. In Java, this is done by maintaining a series of hill gardens which supply cuttings to the factories. The expense is considerable, as cane sets are bulky and costly to transport. If the cause of this gradual deterioration could be discovered, the efficiency of new varieties might be prolonged at much less cost. A closely allied question is the development of sports or mutations in the cane, a phenomenon which is particularly noticeable in striped varieties. Does this occur in all varieties or is it limited to certain types? Are there any soil or climatic conditions which favour the production of mutations?

In the actual cultivation of the cane in India, the chief requirements are combined nitrogen, moisture, and adequate soil-aeration. It is fortunate that a large amount of successful work has already been done on the solution of the nitrogen problem at two of the sugar stations in India—Shahjahanpur in the plains and Manjri near Poona. At Shahjahanpur, an addition of 100 lb. of nitrogen to the acre, in the form of organic manure once in four years, in combination with irrigation and surface-drainage, has raised the production of sugar about threefold besides giving largely increased crops of wheat, gram, and fodder in the succeeding years of the rotation. At Manjri, heavy crops of cane are raised on green manure, supplemented by top dressings of organic matter containing 75 lb. of nitrogen to the acre. Much remains to be done, however, in working out the minimum application of combined nitrogen needed to *maintain* the fertility of cane land once the production has been raised to the Shahjahanpur standard. Provided attention is paid to surface-drainage and to soil-aeration, there is every reason to expect that the increased content of organic matter in the soil not only promotes nitrification but also leads to a larger amount of fixation.

When a nitrogen balance sheet has been worked out for sugar-cane in the plains and on the black soils, it will be possible to lay down a scientific scheme of manuring. While everything points to organic matter like oil-cake as the ideal method of adding nitrogen, nevertheless the present supplies are limited and moreover are not capable of indefinite expansion. If the intensive cultivation of cane is to spread to any great extent, other sources of organic matter must be discovered or an effort must be made to supplement existing supplies by means of synthetic nitrogen obtained from the atmosphere. In Java, sulphate of ammonia is applied to the cane in large quantities with great advantage, but up to the present this substance has not been taken up by the people of India to any considerable extent. Most of the sulphate of ammonia produced in our coal-fields is exported to Java.

Beyond the work done at Manjri and Shahjahanpur, practically nothing is known in India as to the real water requirements of the cane beyond the fact that the cultivators habitually overwater. The Manjri results have shown that cane can be grown on less water than is used by the people. At Shahjahanpur, cane grown on the Java method need not be watered more than three or four times during the growth season. Even these applications may be more than is really required. The matter needs far more investigation. If the cane can be induced to root as deeply as possible in the hot weather, less water would be needed. The water requirements of the cane, the root-development of the variety, and the soil type should all be studied together as parts of a single problem.

## BIBLIOGRAPHY


Abel, P.—Notes on Sugar machinery and manufacture in Northern India, *Bull. 47, Agr. Research Institute, Pusa*, 1915.

Annett, H. E.—The experimental error in field trials with sugar-cane and the effect of this error on various methods of sampling, *Bull. 49, Agr. Research Institute, Pusa*, 1915.

Barber, C. A.—The Samalkota Sugar-cane farm, *Agr. Jour. of India*, i, 1906, p. 44.

——————The production of early-maturing canes, *Agr. Jour. of India*, i, 1906, p. 226.

Barber, C. A.—The origin of new sugar-canes by bud-variation, *Agr. Jour. of India*, i, 1906, p. 285.

————Sugar-cane cultivation in the irrigated lands of the Godaveri delta, *Agr. Jour. of India*, ii, 1907, p. 53.

————Seedling canes in India, *Agr. Jour. of India*, vii, 1912, p. 317.

————Sugar and sugar-cane, *Agr. Jour. of India*, x, 1915, p. 237.

————The classification of indigenous Indian canes, *Agr. Jour. of India*, xi, 1916, p. 371.

————Studies in Indian sugar-canes, Nos. 1–5, *Mem. of the Dept. of Agr. in India* (*Botanical Series*), vii, 1915, p. 1; viii, 1916, p. 102; ix, 1917, p. 133; x, 1919, pp. 39 and 155.

Barnes, J. H.—The after-ripening of cane. Chemical changes which take place after cutting, *Agr. Jour of India*, xii, 1917, p. 200.

————Sugar and the sugar-cane in the Gurdaspur District, *Bull. 69, Agr. Research Institute, Pusa*, 1918.

Brown, W. R.—Paunda sugar-cane cultivation around Peshawar, *Agr. Jour. of India*, vii, 1912, p. 34.

Brown, W. R., Harrison, W. H., and Sanyal, P. B.—Windrowing sugar-cane in the North-West Frontier Province, *Mem. of the Dept. of Agr. in India* (*Chemical Series*), v, 1920, p. 237.

Burt, B. C.—The sugar factory exhibited in the Agricultural Court, Allahabad Exhibition, *Agr. Jour. of India*, vi, 1911, p. 267.

Butler, E. J.—The selection of sugar-cane cuttings, *Agr. Jour. of India*, ii, 1907, p. 193.

Chatterton, A.—The manufacture of jaggery in South India, *Bull. 55, Agr. Research Institute, Pusa*, 1915.

Clarke, G.—Improved sugar-canes in the United Provinces, *Agr. Jour. of India*, xi, 1916, p. 243.

Clarke, G., and Hadi, S. M.—Sugar-cane at the Partabgarh Experiment Station, *Bull. 13, Agr. Research Institute, Pusa*, 1908.

Clarke, G., and Banerjee, S. C.—The efficiency of the Hadi process of sugar manufacture, *Agr. Jour. of India*, v, 1910, p. 29.

Clarke, G., Annett, H. E., and Husain, S. Z.—Experiments on the cultivation of sugar-cane at the Partabgarh Experiment Station 1909–11, *Bull. 27, Agr. Research Institute, Pusa*, 1911.

Clarke, G., Naib Husain, M., and Banerjee, S. C.—Notes on cane crushing in the United Provinces, *Bull. 42, Agr. Research Institute, Pusa*, 1914.

————Notes on improved methods of cane cultivation, Allahabad, 1919.

————Coimbatore sugar-cane seedlings in the United Provinces, *Agr. Jour. of India*, xvii, 1922, p. 445.

Clarke, G., Banerjee, S. C., Naib Husain, M., and Qayum, A.—

Nitrate fluctuation in the Gangetic alluvium and some aspects of the nitrogen problem in India, *Agr. Jour. of India*, xvii, 1922, p. 463.

Coventry, B.—The present position of Indian sugar industry with some suggestions for its improvement, *Inter. Sugar Jour.*, xiv, 1922, p. 30.

Howard, A.—A suggested improvement in sugar-cane cultivation in the Indo-Gangetic plain, *Agr. Jour. of India*, vii, 1912, p. 41.

————The botanical aspect of the improvement of sugar-cane in India, *Inter. Sugar Jour.*, xiv, 1912, p. 35.

Hulme, W.—Sugar production in the United Provinces from an Engineer's point of view, *Agr. Jour. of India*, x, 1915, p. 54.

Hulme, W., and Sanghi, R. P.—The improvement of the indigenous methods of *gur* and sugar making in the United Provinces, *Bull. 82, Agr. Research Institute, Pusa*, 1918.

Leather, J. W.—Manuring sugar-cane, *Agr. Jour. of India*, i, 1906, p. 13.

————Sugar-cane in India, *Agr. Jour. of India*, vi, 1911, p. 255.

————The experimental error in sampling sugar-cane, *Mem. of the Dept. of Agr. in India (Chemical Series)*, iii, 1913, p. 107.

Lehmann, A.—Improvement of the sugar industry of Mysore, *Agr. Jour. of India*, ii, 1907, p. 54.

Mackay, C. J.—Sugar growing and manufacture in Northern India, *Agr. Jour. of India*, iv, 1909, p. 149.

McGlashan, J., and Clouston, D.—The *gur* industry in the Central Provinces, *Agr. Jour. of India*, x, 1915, p. 215.

Moreland, W. H.—The sugar industry in the United Provinces, *Agr. Jour. of India*, ii, 1907, p. 15.

Parr, A. E.—Sugar-cane experiments at the Aligarh Experimental Farm, *Agr. Jour. of India*, vii, 1912, p. 301.

Pillai, N. K.—Some sugar-cane experiments in Travancore, *Agr. Jour. of India*, xi, 1916, p. 79.

Proceedings of the Board of Agriculture in India, Calcutta, 1912.

Progress of the sugar-cane industry in India during the years 1916 and 1917, *Bull. 83, Agr. Research Institute, Pusa*, 1918.

Report of the Indian Sugar Committee, Simla, 1921.

Sayer, W.—The Indian sugar industry, *Agr. Jour. of India*, xi, 1916, p. 42.

————Sugar-cane cultivation in non-tropical parts of India, *Agr. Jour. of India*, 1916, p. 360.

————The problem of sugar manufacture in India, *Agr. Jour. of India*, xii, 1917, p. 550.

————Mill trials of selected Coimbatore sugar-cane seedlings, *Agr. Jour. of India*, xviii, 1923, p. 249.

Taylor, C. S.—Notes on experiments with sugar-cane at Sabour, *Bull. 37, Agr. Research Institute, Pusa*, 1913.

Venkatraman, T. S.—A preliminary note on the behaviour in North India of the first batch of sugar-cane seedlings distributed from the sugar-cane breeding station, Coimbatore, *Bull. 94, Agr. Research Institute, Pusa*, 1920.

——————A study of the arrowing (flowering) in the sugar-cane with special reference to selfing and crossing operations, *Agr. Jour. of India (Indian Science Congress Number)*, xii, 1917, p. 97.

——————Germination and preservation of sugar-cane pollen, *Agr. Jour. of India*, xvii, 1922, p. 127.

Venkatraman, T. S., and Krishnamurti Row.—Study of the sucrose variations in successive cane joints as they attain maturity with special reference to the death of the leaves, *Agr. Jour. of India (Indian Science Congress Number)*, xii, 1917, p. 117.

Venkatraman, T. S., and Thomas, R.—The care and treatment of new sugar-cane importations, *Agr. Jour. of India*, xvi, 1921, p. 24.

——————Sugar-cane root systems: studies in development and anatomy, *Agr. Jour. of India*, xvii, 1922, p. 381.

Wilkie, J. J.—Sugar-cane cultivation in Behar, *Agr. Jour. of India*, i, 1906, p. 322.

Woodhouse, E. J., Basu, S. K., and Taylor, C. S.—The distinguishing characters of sugar-canes cultivated at Sabour with a note on the chemical characters, *Mem. of the Dept. of Agr. in India (Botanical Series)*, vii, 1915, p. 107.

# XIV

# FIBRES

When the present Agricultural Department began its operations in 1905, a good deal of work on fibre plants had already been started by Watt, Prain, Burkill, and Mann. Up to 1911, papers on fibres in the *Agricultural Journal of India* were frequent, but since that date the number has fallen off considerably. With the exception of cotton, the only fibre crops which are now being intensively studied are jute and to some extent the coco-nut palm. Work on the remainder has either been dropped altogether or has become of secondary importance. In this chapter, the opportunity has been taken to review briefly the investigations of the last twenty-five years, to record the advances made, to define some of the problems which await investigation, and to indicate the reasons why certain fibre crops have not been successful under Indian conditions.

*Jute.* One of the most striking developments in Indian agriculture during the last hundred years is the spread of jute cultivation in Bengal. This rapid growth has arisen as a result of the development of manufacturing enterprise in Europe. The industrial population there required large quantities of food; the factories needed abundant material. For both these purposes, seeds had to be transported at the lowest possible cost. Hence the demand for gunny bags for the cheap carriage of seed. Bengal at once responded, and has proved to be the only country in the world capable of producing a suitable fibre cheap enough for the purpose. Since 1828, when the export of jute from Bengal first received commercial mention and when eighteen tons were shipped, the industry has steadily developed to its present enormous dimensions. In 1923, the area under this crop was estimated at 2,313,685 acres and the yield at nearly 7,000,000 bales (each 400 lb.). For a time, the gunny bags exported from India were made in the villages, but after the first jute mill was erected in 1854, the hand-loom industry has steadily lost ground and has now practically disappeared.

Although cultivated over a wide area in Bengal, and to a much less extent in Bihar and Assam, more than half the jute of commerce comes from a comparatively narrow strip of country on both sides of the Brahmaputra in Eastern Bengal. In 1923, the District of Mymensingh alone was estimated to produce nearly 1,750,000 bales—about 25 per cent. of the total annual production of India. After Mymensingh, the chief centres are the Districts of Dacca, Tipperah, Rangpur, Faridpur, and Pabna, all of which lie alongside the Brahmaputra. These six Districts produce the best qualities of jute and about half the total crop. The only other District in which over 100,000 acres are grown is Purneah in North Bihar. The remainder of the crop is mainly produced in the Gangetic delta, where the percentage of the cropped area under jute is much less than in the Brahmaputra region and where the fibre lacks the colour, fineness, and gloss of the product from Eastern Bengal. The superior quality of the jute from the country on both sides of the Brahmaputra appears to be due to two main factors—a comparatively high and well-drained soil and clear water for retting. Although frequently inundated after the rains set in, the jute plant thrives best when adequate soil-aeration is provided. Where this is defective, the plant either forms numerous adventitious roots on the lower part of the stem or becomes chlorotic. A loss of quality and a diminished yield of fibre is the result.

The universal demand for sacks in all parts of the tropics and sub-tropics has naturally led to attempts to start a local jute industry, so as to avoid the importation of gunny bags from Calcutta. Many experiments have been carried out in Cochin China, Japan, Java, Egypt, Russia, the Southern States of North America, and Brazil, but nothing practical has resulted. None of these countries possess all the factors necessary for success, namely (1) a well distributed rainfall throughout the crop, with light showers and ample sunshine in the early stages, (2) a fertile soil, (3) abundant water at the right temperature for retting and for the cheap transport of the fibre, (4) a large population in need of a reliable money crop like jute, and (5) a network of buying and pressing agencies for the rapid disposal of the produce. Serious competition from other countries and from other parts of India is, therefore, not likely to arise. Bengal, for many years to come, need not fear for the monopoly which Nature and the enterprise of her merchants have firmly established.

The natural home of the two species of cultivated jute—*Corchorus capsularis* L., ordinary jute, and *C. olitorius* L., long podded or *daisi* jute—is still a matter of controversy. After an extensive examination of the literature and of the herbarium specimens available at Calcutta and Kew, Watt concluded that it was by no means proved that either of these plants is indigenous to India. He is inclined to believe that they may have been introduced into this country from China. Roxburgh states that a form of ordinary jute with reddish stems, the seeds of which came from Canton, had been successfully acclimatized in Bengal. This, he adds, yields a better fibre than the ordinary Bengal crop. Watt also refers to the use of lime water by the Chinese in the retting of jute, and to a product, superior to the ordinary fibre of commerce, which resembles linen in its general appearance. Hautefeuille, on the other hand, considers that the jute of China is the produce of a species of *Abutilon*. The matter clearly requires further investigation, and a search for the various species of jute grown in China as well as a study of the methods used in retting might easily yield results of value to Bengal. It would be interesting to know whether the well-known susceptibility of commercial jute to damp and other disintegrating agencies is due to an inherent weakness in the fibre or to the methods of retting in vogue in Bengal. If the durability of jute could be improved by a better system of retting, the door might be opened to many new uses for this material.

During the last twenty years, several investigators have devoted attention to the improvement of jute, and it will not be out of place, before defining the problems which still remain to be solved, to sum up briefly the practical results which have been obtained. Two main lines of investigation have been pursued: (1) the cause of heart-damage in the baled fibre, and (2) a survey of the kinds of jute grown with a view to obtaining improved varieties. Heart-damage is the trade name given to the deterioration which sometimes takes place after the jute has been pressed. A large proportion of the fibre in the bale loses its tensible strength and becomes useless for spinning. Finlow has shown that the damage is due to the action of bacteria on the cellulose of the fibre and can only take place if the moisture content has been raised from the normal 10 per cent. to 25 per cent. The remedy is to dry the jute before pressing and to penalize fraudulent watering. The survey of the races of jute grown, which was initiated by Burkill and Finlow, led to the discovery of *Kakya Bombai* in the Serajgunj sub-division of the Pabna District. This variety, originally discovered in a mixed condition, is said to have been introduced into the Pabna District from Goalpara in Assam by a seed dealer. From the original seed, three unit species have been selected at the Dacca Experiment Station which yield more fibre per acre than the mixtures ordinarily grown. These improved types have been named *Kakya Bombai* No. 7, R. 85, and D. 154. The two latter are considered to be the best types and to be more resistant to chlorosis than *Kakya Bombai* No. 7. The seed of R 85 and D 154 is now being distributed to the cultivators. Similar selection work has been carried out in the case of *daisi* jute, and *Chinsurah green* is considered to be the highest yielder of the types investigated. As the amount of cross-fertilization between the unit species of both kinds of jute is small and *Corchorus capsularis* and *C. olitorius* do not cross, the isolation of improved kinds and the raising of seed in bulk are attended by no theoretical difficulties.

In the actual distribution of the seed of the improved varieties to the cultivators, a number of practical difficulties have had to be overcome. Although Bengal is admirably adapted for the production of fibre, it is by no means an ideal locality for the production of jute seed. The high humidity, the continuous rain, and the frequent lack of sunny days interfere with the pollination of the flowers. The water-logged condition of the soil towards the end of the season prevents the proper maturation of the seed. These adverse factors explain the low yields of jute seed obtained in Bengal.

Further, there are no large areas suitable for seed-farms. These circumstances have prevented the Agricultural Department from obtaining a large bulk of seed of the improved kinds with which to replace the country kinds over extensive areas. This difficulty has been met, in part, by growing the improved varieties for seed purposes on certain estates in Bihar and Assam and in the forest nurseries in Bengal, where jute is used as a foster plant for the young trees. The total amount of seed obtained, however, has been relatively small, and this has interfered with the distribution of the new types and with the effective 'follow up' of the work of the Experiment Stations.

The Agricultural Department has naturally striven to make the best possible use of the small supplies of seed at its disposal. With this object, an interesting experiment was made between the years 1916 and 1920. A method of extensive distribution of the seed of the new types was tried. The seed available was broken up into quarter-pound packets and distributed to a large number of cultivators, through the village panchayats, on condition that the resulting crop should be kept for seed the next year, and that the seed given out by the Agricultural Department should be repaid in kind. The system, however, broke down. Many of the cultivators did not carry out their promise to multiply the seed for their next sowings. The seed returned to the Department was often so impure that it was useless for further distribution. This method of extensive seed-distribution has now been given up in favour of a more intensive system, on the lines worked out for wheat in the United Provinces (p. 71). Efforts are being made to enlist various local agencies for seed growing and also to increase the bulk of seed obtained outside Bengal. Excellent jute seed can be grown in tracts like Oudh and Rohilkhand, but the difficulty will be to organize suitable local agencies for the work.

After an effective system of growing, storing, and distributing the seed of improved varieties (suitable for the various jute areas) has been set in motion, the next problem that arises is the increase in the yield per acre by more intensive methods of agriculture. A great deal of the preliminary work in this direction has already been done by the Agricultural Department. There is no doubt that when finely divided organic matter containing a fair proportion of nitrogen is applied to the best jute lands, the plant responds at once by more rapid growth and by an increased yield of fibre. The land is also left in a suitable condition for autumn rice, which often follows jute. The conventional source of organic matter for this purpose is either cow-dung or oil-cake, but the limited supplies of both, as well as the cost of the latter, are great obstacles to progress. Where then is the manure needed to be obtained? Nature has fortunately provided unlimited supplies of the raw material on the spot for the mere cost of collection. The raw material is the water hyacinth (*Eichornia crassipes*, Solms), which analysis shows is rich in organic matter, potash, and also contains a good deal of nitrogen and phosphorus. In its fresh condition, this water-weed is of course unsuitable, but by applying to this material the Chinese methods of composting organic matter, a very valuable manure could probably be obtained. Besides the weed, mud from the low-lying areas where the plant grows, earth, ashes, a little cow-dung, and labour are all that is required. The materials are available on the spot. The Chinese methods have been described by King in his *Farmers of Forty Centuries*. It should not be difficult to adapt them to Bengal conditions.

*Sann-hemp* (*Crotalaria juncea* L.). In spite of the attention which has been paid in the past to the development of this fibre plant, little or no increase in production has taken place in recent years. The crop is grown all over India for domestic purposes, the surplus being exported. Why has no increase in the production of *sann* fibre taken place? The answer becomes apparent when the root-system of the plant is examined. Sann-hemp only thrives on permeable, well-aerated soils; water-logging is harmful for fibre purposes and still more so for seed production. These requirements have helped to drive the cultivation away from the deltas, towards the higher and better drained parts of the country. This is satisfactory as far as growth is concerned, but a difficulty at once arises in retting. In the drier parts which suit the crop, there is apt to be a shortage of water in the autumn for retting.

This circumstance has naturally limited production and, in all probability, accounts for its stationary condition in spite of the excellence of the fibre. The cultivation of sann-hemp is therefore not favoured by circumstances to anything like the same extent as jute.

The fact that the yield and quality of fibre varies with the locality suggests that the crop should receive intensive study with the object of producing higher yielding varieties. One great obstacle, however, stands in the way. The flowers of this plant are self-sterile, and only set seed after being visited by insects. Cross-fertilization is the rule, and the crop is a complex mixture of forms which do not breed true. The labour involved in isolating an improved type and in the organization of the seed-supply would be quite beyond the means of the Agricultural Department. Even if funds were unlimited, it is doubtful whether the improvement possible would be sufficient to justify the expenditure. The only practical improvement, as far as the variety is concerned, is the introduction of seed from other tracts. Even this is not always successful. The introduction of the superior variety from Jubbulpur into the plains, although increasing the yield and quality of fibre, is of no permanent advantage, as this deep-rooted type produces so little seed that acclimatization is impossible.

There is one minor difficulty which in the past has interfered with trade, namely, admixture of the fibre with sand and mud. On several occasions, particularly in the Central Provinces, adulteration has been carried on to such an extent that the intervention of Government has been necessary to put a stop to these practices. This is not likely to be required in the future. The factory regulations in Europe are now becoming more and more effective. Raw products which are adulterated with fine sand and mud are harmful to the workers, and the use of such materials is no longer permitted. Stricter factory regulations are affecting the trade and are leading to the rejection in India itself of all doubtful consignments.

*Patwa* (*Hibiscus cannabinus* L.). This crop is widely grown to supply the fibre needed for the ropes and cord required in the villages. In Northern India, it is often sown as one of the constituents of the mixed monsoon crops. In the Vizagapatam

District of North Madras, however, it occurs as a pure crop; the fibre coming into the trade under the name of Bimlipatam jute. The plant possesses a deep root-system and, for the best results, requires an open, well-drained soil in good condition. It suffers from the same disadvantage as sann-hemp, and cannot therefore compete with jute. An improved type, with a much shallower root-system than the average, has been isolated at Pusa, and seed has been distributed in small quantities. The scattered nature of the cultivation and the fact that crossing readily takes place in this species will make the systematic replacement of the country crop by the new type a matter of considerable difficulty. In South Africa, however, more promising results have been obtained. The new variety has yielded well there, and a Company is being formed to cultivate it on the large scale.

*Flax*. While the foreign trade of India has practically created the present cultivation of linseed as a source of oil, it has led to no developments from the point of view of flax. This result is not due to lack of interest or to failure to make experiments. During the last hundred years, attempt after attempt has been made to grow flax in India and also to combine the production of linseed oil and fibre. Without exception, all the experiments have ended in comparative failure. In the years preceding the Great War, the last of these attempts was made at the Dooriah estate in Bihar, as a result of representations made by the Dundee Chamber of Commerce to the Secretary of State for India. A Belgian flax expert (M. E. Vandekerkhove) was engaged for a period of six years, from March 1907 to April 1913, for the work of retting and preparing the fibre for market. The resources of the Dooriah estate were made available for the growth of the plant. Both European, acclimatized and ordinary Indian seed were tried, and modern methods of retting and skutching were installed. The experiments were conducted with great care, and the results, including the details of cost, were published in Bulletins 25 (1911), 30 (1912), and 35 (1913) of the Pusa Research Institute. During the progress of the work, a number of important facts were ascertained. Although the imported seed grows in Bihar, nevertheless it only does really well on specially selected fields in good condition, and in years when the moisture in the surface soil is adequate for good germination. Years of short rainfall, which occasionally occur in Bihar, and the frequent failure of the sowing rains in early October are distinctly unfavourable circumstances. Acclimatized seed rapidly degenerates, and cannot be sown with safety after the second or third year. The growth of flax, therefore, involves the frequent importation of fresh seed from Europe. As the seed-rate for fibre is high and the cost of the seed is considerable, this would be a very serious item from the cultivator's point of view. The greatest difficulty encountered, however, was in the retting process, which requires a comparatively low temperature. After the crop is pulled at the end of the cold weather, the water soon becomes too warm for retting and a good deal of the flax straw has to be stored in well-made godowns till after the rains, when abundant water at the right temperature is again available. This is a fatal drawback from the point of view of the cultivator who is compelled to sell his money crop at harvest time for cash. These disadvantages, combined with the fact that Indian-grown flax does not command more than moderate prices, have also deterred capitalists from embarking in the industry.

What are the prospects of the ordinary Indian crop for combining fibre and oil? All attempts to produce marketable flax at Dooriah from Indian linseed failed completely. There appears to be some factor like temperature which is operating against the production of really high-quality flax in India. A certain amount of cold appears to be essential for fibre; hot climates seem to favour the production of oil. In view of past experience with flax, therefore, there appears to be no point, in the present state of knowledge, in making any further trials in India. The development of the linseed crop, which is already important, offers much greater chances of success.

*Sisal hemp.* In the early years of the present century, a good deal was written about the prospects of sisal hemp in India. Several plantations were started about that time, on certain tea estates in Assam, in Chota Nagpur, and in the United Provinces. The species most suited to Indian conditions was found to be *Agava sisalana*, Perrine. The conditions necessary for success were studied in detail by Mann, when Scientific Officer to the Indian Tea Association. The results of his investigations were published in 1904 in a pamphlet entitled *Sisal Hemp culture in the Indian Tea Districts*. The plant did best on good well-drained land, where the rainfall is well distributed, so that there is no check in growth. Suckers gave better results than bulbils. The cultivation is only suitable for the capitalist in areas of India like Assam, where suitable land is available at a small cost and where there is sufficient rainfall. Since 1906, practically nothing has been published on the subject. The cultivation has not been taken up by the people, while the area on the Assam estates has greatly diminished. There is, however, one locality in the Central Provinces where sisal is grown on a large scale. In the Kawadha Feudatory State adjoining Bilaspur, the fibre is made into ropes and cloth. The crop does not appear to be popular, as the juice of the leaves produces eczema on the arms and legs of those engaged in the extraction of the fibre.

*Coir.* On the Malabar and Coromandel Coasts of the Madras Presidency, the coco-nut is widely cultivated both for fibre (coir) and for oil. The produce of the Laccadive and Maldive Islands also finds its way to South Indian ports and helps to swell the total exports. For some years past, attention has been paid to the coco-nut by the Agricultural Departments of Madras and Travancore, and a number of farms have been opened for the experimental study of this tree. The results have recently been brought together by Sampson in *The Coco-nut Palm*, a work which should be studied by all interested in this subject. One of the features of this book is the study of the root-system and its relation to the aeration of the soil. Land with free drainage is perhaps the chief condition of success in the cultivation of this crop.

*Rhea or China grass.* Towards the end of the nineteenth century, many attempts were made in India to establish a *rhea* industry. For this purpose, a number of companies and plantations were started, but all of these have disappeared. The various experiments made and the results obtained are summed up in *The Commercial Products of India*, and should be carefully considered before anything further is attempted. The industry appears to have failed for two reasons. In the first place, the plant is very exacting in its requirements and only thrives on permeable, well-aerated soils in excellent condition where the rainfall is adequate and well distributed. Such soils are already in great demand for money crops like tobacco, chillies, sugar-cane, and vegetables, in which the yield is assured and for which a remunerative market already exists. Very powerful inducements, therefore, will be needed before the cultivator will adopt a new crop on such lands. In the second place, the prices realized for the fibre were not high enough to cover the cost of cultivation and of extraction. There is nothing, therefore, in *rhea* to attract either the cultivator or the capitalist. In spite of the fact that the plant will grow in India, that numerous trials have been made, and that the fibre possesses great strength and durability, the cultivators generally have not adopted the crop even as a source of domestic fibre. It appears to be grown in small quantities in Bengal for the manufacture of fishing-nets. Beyond this, the plant has failed to find a place in the rural economy of India.

## BIBLIOGRAPHY

Administration Reports of the Department of Agriculture, Bengal, Calcutta, 1916–23.

Burkill, I. H., and Finlow, R. S.—The races of jute, *Agricultural Ledger*, 1907, p. 41.

Burt, B. C.—Flax in the United Provinces, *Agr. Jour. of India*, xv, 1920, p. 616.

Butler, E. J.—Report on Coco-nut palm diseases in Travancore, *Bull. 9, Agr. Research Institute, Pusa*, 1908.

Chaudhury, N. C.—Jute in Bengal, Calcutta, 1921.

Clouston, D.—Sann-hemp, Ambari, and Agave as fibre crops in the Central Provinces and Berar, *Agr. Jour. of India*, iii, 1908, p. 144.

Coventry, B.—Rhea experiments in India, *Agr. Jour. of India*, ii, 1907, p. 1.

d'Lima, C.—Indian hemp fibre, *Agr. Jour. of India*, xi, 1916, p. 31.

Finlow, R. S.—The extension of jute cultivation in India, *Bull. 3, Agr. Research Institute, Pusa*, 1906.

——————The improvement of the jute crop by pure line selection, *Agr. Jour. of India*, xii, 1917, p. 283.

Finlow, R. S., and McLean, K.—Water hyacinth (*Eichornia crassipes*), its value as a fertilizer, *Bull. 71, Agr. Research Institute, Pusa*, 1917.

Finlow, R. S.—' Heart damage ' in baled jute, *Mem. of the Dept. of Agr. in India (Chemical Series)*, v, 1918, p. 33.

————————Historical note on experiments with jute in Bengal, *Agr. Jour. of India*, xvi, 1921, p. 265.

Fruwirth, C.—Handbuch der landw. Pflanzenzüchtung, Bd. v, Berlin, 1923.

Hautefeuille, L.—La culture du Jute dans l'Inde et en Indo-Chine, *Bull. Economique de L'Indo-Chine*, Nos. 113–14, 1915.

Howard, A., and Howard, G. L. C.—An improved fibre plant, *Agr. Jour. of India*, x, 1915, p. 224.

Howard, A., Howard, G. L. C., and Abdur Rahman Khan.—Studies in the pollination of Indian crops, I, *Mem. of the Dept. of Agr. in India (Botanical Series)*, x, 1919, p. 195.

Mann, H. H.—The development of sisal hemp cultivation in India, *Agr. Jour. of India*, ii, 1907, p. 323.

Mann, H. H., and Hunter, J.—Sisal hemp cultivation in the Indian Tea Districts, Calcutta, 1904.

Note on the extension of cultivation of fibre plants in India, *Bull. 15, Agr. Research Institute, Pusa*, 1909.

Sampson, H. C.—The coco-nut palm. The science and practice of coco-nut cultivation, London, 1923.

Vandekerkhove, E. M.—Reports on the flax experiments conducted at Dooriah, *Bulls. 25, 30, and 35, Agr. Research Institute, Pusa*, 1911–13.

Watt, G.—The Commercial Products of India, London, 1908.

# XV

## OIL-SEEDS

Oil-seeds are an important item in the rural economy of India and constitute a valuable group of money crops. The area under the five principal oil-seeds—linseed, rape, mustard, sesamum, and ground-nut—was, in 1922, about 16,000,000 acres and the yield a little over 3,000,000 tons. The export during the year was 735,000 tons, valued at 1,746 lakhs of rupees. This is about half the pre-war average export and a quarter of the total produce. Besides the five crops mentioned above, the production of and trade in the minor oil-seeds, such as gingelly, poppy, and *mahwa* (*Bassia latifolia* Roxb.), is considerable, while the export of copra and cotton seed—the produce of crops not solely grown for oil—is an important commercial item. During the pre-war period, when the export trade was at its maximum, about half the oil-seeds produced were used up in India itself for purposes of internal trade and for domestic use in the villages. Besides the shipments of seeds, considerable quantities of coco-nut and linseed oil are exported from Madras and Calcutta respectively, while a good deal of oil-cake also finds its way overseas.

The export trade in oil-seeds and in oil-cake, with its resulting loss of nitrogen to the country, is a matter which periodically comes up for discussion in India. Interest was re-awakened in oil-crushing as a result of the *Report of the Indian Industrial Commission*, published in 1918. The loss of soil fertility resulting from the export trade was discussed by the Board of Agriculture in 1919 and again in 1922. There is no doubt that the export of oil-seeds and oil-cake represents a serious loss of valuable manure to the country, and that if it could be obviated, rural India would be the gainer. On the other hand, it must not be forgotten that oil-seeds provide a very useful source of income for the cultivator. From the point of view of the loss of nitrogen, the position is by no means so serious as it is often represented. It is sometimes forgotten that the combined nitrogen lost in the shipments of oil-seeds is automatically replaced by fixation of free nitrogen from the atmosphere. If this were not the case, the soil of India would have been exhausted long ago, and the traffic in these and similar products would have come to an end. After making every allowance for the benefits derived from nitrogen fixation, the fact remains that it would be a great help to Indian agriculture if more oil-cake were available for such crops as cotton, tobacco, sugar-cane, and tea. The only means of accomplishing this is to crush the seeds in India, to export the oil, and to sell the cake to the cultivators. The adoption of this remedy is attended by several difficulties. If oil-seeds are pressed in India, practically all the oil will have to be exported to the manufacturing areas of Europe, as very little can be utilized in India itself in the present state of industrial development. This involves the transport of oil without deterioration or loss and its sale in the industrial centres of Europe. Oil-seeds can be moved cheaply in gunny bags by means of the existing transport agencies. The sacks find a ready sale afterwards. The movement of oil is another thing. Expensive metal has to take the place of cheap gunny. The freight on oil is much higher (and more difficult to obtain) than on seeds. Questions of leakage and deterioration arise, while oil-drums are not so easily handled or stored as sacks of seed. Even if the problem of the cheap transport of oil to Europe could be solved, other difficulties remain. The European industries engaged in crushing oil-seeds and in utilizing oils must also be considered. They are not likely to remain passive spectators while drastic changes in the trade are in progress. How would these interests view the stoppage or curtailment of the present export of oil-seeds from India? It has always been tacitly assumed that if India were to export vegetable oils instead of oil-seeds, the oil would find as ready a market as the seeds do now. But would it? Modern industries demand a uniform and reliable product. The users of oil are not likely to forsake their present supplies in favour of a distant source unless there is a distinct saving in the cost of the raw material and unless the new product proves to be reliable. The oil-pressing industries would be certain to strain every nerve to maintain themselves and also to preserve the trade in oil-cake, which is as essential to agriculture in Europe as oil is to her industries. Improvements are constantly taking place in oil-pressing, and India might easily be left behind in the race. Moreover, this country holds no monopoly in the production of oil-seeds. During the last fifteen years, India has had to take the second place in the world's supply of linseed. The Argentine, which at the beginning of the century was only beginning to export this seed, now heads the list. Cotton seed, formerly a waste product, has now become one of the most important oil-seeds of commerce. Africa, which is being rapidly developed, already produces large quantities of oil-seeds and could grow many, if not all, of the products now furnished by India. China is already a formidable rival in the production of sesamum. Roumania has become a serious competitor in rape seed. It by no means follows, therefore, that the establishment of an oil-pressing industry in India would help this country. On the contrary, it might deflect the present profitable trade in raw materials to Africa and elsewhere and do permanent harm to the cultivators.

Besides the disposal of the oil at remunerative prices, a large amount of oil-cake would remain for sale in India. Some of this would have to be fed to animals, the rest used as manure. Unless most of this cake is used up in India, the agricultural advantage of a large indigenous oil-crushing industry would not be realized. At the present time, very little of the oil-cake now produced finds its way to the land, as the prices commanded by this product on the open market are beyond the means of the average cultivator. In consequence, large quantities have to be exported. Increasing the supply, under present conditions, would have no effect on the amount absorbed locally. The supply is already greater than the local demand.

The prospects of establishing a large oil-crushing industry in India at the present time under free trade conditions are, therefore, not very bright. To assist the new industry, two measures have been suggested—the prohibition of the export of oil-seeds altogether and the imposition of an export duty on oil-cake. The first of these proposals would merely interfere with the balance of trade to the detriment of India. An export duty on oil-cake, as pointed out in the discussion of this subject at the 1922 meeting of the Board of Agriculture, would lead to a corresponding reduction in the market price in India and the cultivator would be saddled with an additional burden. Everything, therefore, points to a policy of non-interference with the existing trade, until the industries of India have developed sufficiently to absorb a large quantity of the oil extracted.

Another aspect of the production of oil-seeds in India has almost escaped notice in recent discussions, namely, the competition of mineral oils for burning and lubrication. The advent of cheap kerosene, put up in a handy and readily saleable package, combined with the manufacture of cheap lamps, has greatly reduced the internal consumption of vegetable oils (such as castor, coco-nut, and colza) for lighting purposes, and must have had a great effect on the internal trade in oil-seeds.

A good beginning has been made by the Agricultural Department in the improvement of oil-seeds, and sufficient has been accomplished to indicate the main lines of future progress. Except to a limited extent in linseed and possibly in the castor crop, questions of quality hardly arise in the improvement of these crops. The chief aim of the investigator is to increase the yield and to reduce the cost of production. This greatly simplifies the problem.

In the case of *sarson*, *rai*, and *toria*—the rape and mustard seed of commerce—gingelly, safflower, and castor, the amount of natural cross-fertilization is so great that the labour involved in isolating high-yielding unit species and in maintaining them in pure culture is greater than the means of the present Agricultural Department allow. Further, in several of these crops, a good deal can be said in favour of a judicious mixture of types and of free inter-crossing. This would maintain the vegetative vigour and also prolong the period of pollination, thereby ensuring the setting of a full crop of seed. In rape and mustard, for example, rainy weather at the time of flowering interferes with fertilization. A longer flowering period than is possible with a single unit species is, therefore, an undoubted advantage. In these crops and probably also in sesamum, selection on too narrow a basis is not likely to prove advantageous. Form-separation, within rather wide limits, which would allow of crossing between the types, is perhaps the utmost the plant breeder can accomplish. After this, attention should be paid to increasing the yield by simple methods of manuring, such as the ploughing in of green crops, and the improvement of the soil by the use of composts. In the case of the castor crop, which thrives best on sandy soils as a mixed crop, conditions which favour the production of an oil free from acidity, and suitable for the lubrication of aeroplanes and airships, is a problem worthy of investigation. The question to be settled is whether the slight acidity in castor oil is due to the variety, to the environment, or both, and what is the easiest way of removing it. A beginning has been made by Somers Taylor at Sabour in investigating the relations between oil-content, the variety and the environment, but unfortunately the work has come to an end. The formation of oil in plants like castor appears to be influenced more by environment than by the variety, and seems to depend among other factors on adequate soil-aeration. Another problem presented by this crop is the reason why it is so liable to insect attack when grown in pure culture. On many soils on which castor grows well as a mixed crop, its extension as a pure crop is often followed by defoliation by caterpillars. The cause of this extreme susceptibility might throw light on the mineral requirements of this plant.

In the improvement of linseed and ground-nuts there is much more scope for the plant breeder. In both these crops, self-fertilization is the rule. Further, the range of types is very great. More than a hundred unit species of Indian linseed, which differ considerably in yielding power, have been isolated and studied at Pusa. The types fall into three main classes, depending on the size of the seed and the character of the root-system. The linseeds from the deep, black soils of the Peninsula have large bold seeds and deep root-systems which suit the soil type. These forms ripen early and are admirably adapted to make the most of the available moisture in the black soils. They vary, however, in yielding power and probably also somewhat in the distribution of their roots, so that there appears to be some scope for the breeder in fitting the right kind of plant to the different soil types. On the black soils, linseed sickness occurs and it is said that the plant cannot be grown successfully except once in five or six years. The nature of this soil sickness needs investigation and experiments are desirable to discover whether the unsuitability of the root-system is concerned with the real cause. The kinds of linseed found on the alluvium are quite distinct from the large-seeded forms of the Peninsula. They are small-seeded, shallow-rooted, and relatively late. The yield, however, is much larger than that given by the large-seeded kinds. On the transition soils, which occur near the line of the Jumna, a set of types intermediate in the size of seed and in root-development occur. Natural selection, therefore, has had its usual effect on the botanical composition of the crop, and the types found in cultivation fit the soil type. Botanical selection, however, brings about a better gear between the soil and the plant. Thus several of the small-seeded types isolated at Pusa yield at least 40 per cent. more seed than the mixtures now grown by the people. Seed of these improved types is now being tested in Northern India at the experimental farms and by selected cultivators. If the large seeds of the Peninsular types could be combined with the shallow root-development of the plains, a still further increase in yield might be obtained. Work in this direction is in progress at Pusa. At Nagpur, hybridization experiments are being carried out with the object of obtaining a high-yielding type with yellow seed. As natural cross-fertilization occurs to some extent in Indian linseed, it is important, in all seed-distribution schemes, that the improved variety should possess some distinctive character (such as corolla colour) which separates it easily from the country crop.

The ground-nut presents a number of points of interest. The crop thrives best on well-drained, open soils where the rainfall is well distributed. Opinions differ as to the effect of the ground-nut on the nitrogen supply. On the black soils, it is held to increase the fertility ; on sandy soils, manuring is considered necessary for a full crop. Possibly a part of the beneficial results which follow a crop of ground-nut on the black soils is due to the thorough aeration brought about by the harvesting of the pods. The precise effect on the nitrogen supply is a matter which needs careful investigation, as well as the relation between root-development and the soil-type. It is possible that the *tikka* disease of ground-nuts, which did so much damage in the early part of the century in Bombay and Madras, was more a consequence of unsuitable root-development than a real disease. The disease, however, so reduced the yield that energetic steps had to be taken to save the crop. A number of foreign varieties from America, Europe, and other countries were introduced and tested. The results were singularly successful. Not only was the industry saved, but a large extension of cultivation has taken place in Madras, Burma, and in Bombay—the chief centres of production. The crop has also spread to new areas—Bundelkhand, Orissa, and the Central Provinces. The extension of ground-nut cultivation in India is one of the notable achievements of the Agricultural Department, and no pains should be spared to follow up the pioneer work by an intensive study of the crop and by the isolation of varieties still more suited to the various soil types and to the growth periods of the various tracts. There are many forms in this species, ranging from early, upright types, with all the pods near the base of the plant, to late, spreading types of high potential yielding power. Self-fertilization is the rule in the ground-nut. Consequently, great possibilities of improvement exist. One important problem in several tracts is to find a variety which will ripen rapidly and which can be harvested before the soil dries. Attention to this point, in the Central Provinces, has enabled ground-nuts to be followed by wheat. In Madras, a special farm for the study of this crop has been started at Palur, where selection work is in progress.

Among the minor oil-seed crops of India, safflower is one which might well repay further investigation on the part of the industrial chemists of Europe. At present the seed is crushed in India and the oil is consumed locally, the cake being used for manuring crops like sugar-cane. In many parts, the oil is one of the chief adulterants of *ghi* (clarified butter). The commercial possibilities of safflower seed are dealt with in Bulletin 124 of the Pusa Research Institute, in which the results of a technical investigation of this seed are published. Safflower oil, if properly manipulated, is likely to prove of use in the colour, paint, and varnish industries, for soap and linoleum manufacture, as well as for edible and culinary purposes. The cake has a remarkable effect on the physical properties of heavy soils, and therefore possesses a value in excess of that suggested by its chemical analysis. When the commercial and agricultural possibilities of this seed are realized in Europe, an increasing demand is likely to arise. The crop is easily grown and is well suited to many of the drier soils of India, including the Bombay Deccan.

Perhaps the most interesting development in the oil-seed trade of India in recent times is the export of cotton seed. For a long time this was regarded as a waste product and only fit for cattle food. Little or no use was made of the oil. Various cotton-seed oil products have, however, become important articles of food. Others are used in the production of illuminants, lubricating oils, and in the manufacture of soap. The rapidity with which this seed, so lately looked upon as useless, has been absorbed by industry is of great interest from the point of view of the production of oil-seeds in India. At any moment discoveries in applied chemistry in Europe and America may affect the demand and consequently the price of the oil-seeds at present cultivated. The rapid opening up of Africa is another important factor. In the future, a great deal will depend on an increase in the yield of oil-seeds in India, so that the cost of the raw material can be kept within reasonable limits. Nothing stimulates the inventor so much as high prices and shortage of raw material. Nothing seems to safeguard existing trade so much as ample supplies at a moderate cost.

## BIBLIOGRAPHY

Butler, E. J.—Report on Coco-nut palm disease in Travancore, *Bull.* 9, *Agr. Research Institute, Pusa*, 1908.

————————*Tikka* disease and the introduction of exotic groundnuts in the Bombay Presidency, *Agr. Jour. of India*, ix, 1914, p. 59.

Fruwirth, C.—Handbuch der landw. Pflanzenzüchtung, Bd. 5, Berlin, 1923.

Graham, R. J. D., and Roy, S. C.—Linseed (*Linum usitatissimum*) hybrids, *Agr. Jour. of India*, xix, 1924, p. 28.

Howard, A., Howard, G. L. C., and Abdur Rahman Khan.—Studies in Indian Oilseeds, No. 1, Safflower and Mustard, *Mem. of the Dept. of Agr. in India* (*Botanical Series*), vii, 1915, p. 237.

Howard, A., Howard, G. L. C., and Abdur Rahman Khan.—Studies in the pollination of Indian crops, I, *Mem. of the Dept. of Agr. in India* (*Botanical Series*), x, 1919, p. 195.

Howard, A., and Remington, J. S.—Safflower oil, *Bull. 124, Agr. Research Institute, Pusa*, 1921.

Howard, G. L. C., and Abdur Rahman Khan.—Studies in Indian oilseeds, No. 2, Linseed, *Mem. of the Dept. of Agr. in India* (*Botanical Series*), xii, 1924, p. 135.

Kelkar, G. K.—Ground-nuts in the Bombay Deccan, *Bull. 41, Dept. of Agr., Bombay*, 1911.

Kunjan Pillai, N.—Coco-nut, the wealth of Travancore, *Agr. Jour. of India*, xiv, 1919, p. 608.

McKerral, A.—Burmese sesamum varieties, *Jour. of the Asiatic Soc. of Bengal*, xiv, n.s., 1918, p. 39.

Proceedings of the Board of Agriculture in India, 1919 and 1922, Calcutta, 1920 and 1922.

Report of the Indian Industrial Commission, Calcutta, 1918.

Sampson, H. C.—The Coco-nut palm, London, 1923.

Taylor, C. S.—Experiments with castor seed conducted at Sabour, *Bull. 117, Agr. Research Institute, Pusa*, 1921.

Watt, G.—The Commercial Products of India, London, 1908.

# XVI

# TOBACCO

After the introduction of the tobacco plant into India by the Portuguese more than 300 years ago, the cultivation of this crop spread rapidly all over the country, and at the present time the area is over a million acres. Taking the average value of the crop at 100 rupees an acre, the total cultivation represents a product worth about ten crores of rupees a year. Tobacco is, therefore, an important money crop, and is only grown by the best cultivators. The cured product is always a valuable article of trade in the localities where the area under tobacco is considerable. Two species are cultivated in India. In Madras, Bombay, Burma, Bengal, and Bihar, where the climate is both warm and moist, ordinary tobacco, *Nicotiana Tabacum* L., is grown. In the drier, colder regions of North-West India where irrigation is essential or where, as in Eastern Bengal, the growth period is shortened by the late subsidence of the rain inundation, *Nicotiana rustica* L., a robust, yellow-flowered species with a short growing period, predominates.

Although tobacco is met with all over India, the crop only becomes important in tracts where the texture of the soil is sufficiently open for rapid root-development to take place and where the supply of soil moisture is adequate. Besides a suitable soil texture and sufficient moisture, tobacco requires an ample supply of combined nitrogen. This is usually supplied in the form of easily nitrifiable organic matter, such as cow-dung or indigo refuse (*seeth*). In some cases the nitrogen and other minerals required are applied dissolved in water. In various parts of India, particularly on the sites of old cities in the United Provinces, wells containing the necessary nitrates and other salts in solution (in amounts suitable for the growth of tobacco) occur in conjunction with open, permeable soils. On such lands, as many as three crops, one of which is yellow-flowered tobacco, are grown every year. In these localities, the quality of the cured tobacco is far above the average, and the produce is sold in the large cities at a substantial premium. Tobacco is admirably suited for intensive cultivation. It is a certain crop, and complete failure is rare. Growth is rapid, the yield is large, and there is a steady local demand at good prices. With the exception of a root parasite (*Orobanche cernua* Loeffl.) the crop is remarkably free from insect and fungoid pests, a result perhaps due to the fact that the cultivation is expensive and in the course of time has become restricted to soils which really suit the plant.

Indian tobacco is usually sun-cured, the process frequently being carried out on the ground. This is always the case in the colder parts of India, where yield is the chief aim of the grower. More elaborate curing methods are only found in the damper parts, such as the cigar-filler areas of Bengal and Madras. Nowhere, however, has anything been developed approaching the modern barn or flue curing methods of the United States. Wrapper leaves can only be grown successfully in certain favoured localities, such as the Rungpur District of Bengal and in parts of Madras. The tobacco belt of Bengal furnishes the fillers of the Burma cheroots, while in the neighbourhood of Dindigul and a few other centres in South India, cigars are manufactured for local use and also for export. With these exceptions, practically the whole of the Indian crop is consumed locally, partly for the *hookah* and partly for *snuff*. An important change, however, has taken place during the last twenty years. Cigarettes are now replacing the *hookah*, and a number of factories have been established in India to supply the growing demand. These factories draw their supplies of raw material entirely from India itself, and their operations have created a demand for cured leaf with a texture above the average.

Indian tobacco has not escaped the attention of the improver. Since 1829, the old records abound in efforts to improve and to develop the cultivation of tobacco. Seed has been introduced over and over again from Havana, Virginia, and Sumatra. Many expert curers have been engaged to teach the people better methods of dealing with their produce. Two Government estates—Pusa in old Bengal and Ghazipur in the United Provinces—were leased to a company for developing the industry on modern lines. Attempts have also been made to grow wrapper leaves under shade, according to the system in vogue in the United States. An interesting account of these earlier experiments, all of which failed, is to be found in the fifth volume of the *Dictionary of Economic Products of India*. Want of success appears to have been due to attempting too much. On the one hand, insufficient attention was paid to the requirements of the European market, and, on the other hand, to the conditions under which tobacco is grown and cured in India. As is well known, there is a heavy duty in European countries both on unmanufactured (raw) tobacco and on the finished product. In Great Britain, the duty is much higher on manufactured tobacco than on unmanufactured, a circumstance which favours the importation of the raw material. Further, the leaf imported must be uniform and true to grade from year to year, free from sand and dust, and without stalk. The moisture content must be as low as possible, and no variation in this respect is permissible. While it was quite possible, under experimental conditions, to produce grades of leaf suited to the English market, difficulties at once arose in ensuring regular supplies, true to grade, when the area was extended and expert supervision became less intense. The cultivator could not compete with the United States, and so India failed to establish her position in the Home markets. The circumstances under which tobacco is grown, cured, and marketed in India will always make competition with America a matter of great difficulty. The holdings of the individual growers are small, and their means are quite insufficient for the provision of even the simplest curing barn, constructed on modern principles. An Indian tobacco field rarely ripens off at one time, so that evenness in the cured product is out of the question. This want of uniformity in the crop, before curing begins, places India at a great disadvantage. In addition, dry winds and dust storms follow the growth period so closely that a lengthy curing process is impossible. The cultivators are so intent on yield that coarse varieties are the rule, and as much of the stalk as possible is always included. Little attention is paid to evenness, to colour or to texture. Everything is done to secure the maximum weight, consequently the coarser kinds usually predominate. There is little wonder, therefore, that the ambitious schemes of the last century failed and that the tobacco factories at Pusa and Ghazipur had to be closed down.

The investigations on Indian tobacco which have been carried out by the present Agricultural Department have been on a much more modest scale than the earlier experiments. Care has been taken to limit the scope of the work to simple improvements within the means of the growers. The main lines of the work have been fourfold: (1) the isolation of the various types found in the Indian crop, (2) the reduction in the cost of production by the substitution of green manure for cow-dung, (3) the search for a type of tobacco suitable for the manufacture of cigarettes, and (4) the development of an indigenous cheroot industry in the Rungpur District of Bengal. Twenty different types of yellow-flowered and fifty-one kinds of ordinary tobacco have been isolated at Pusa, and work is in progress in testing these for yielding power. At several centres it has been shown that the amount of manure needed by tobacco can be greatly reduced by green-manuring, provided this operation is completed in time and the surface-drainage is good. The search for a suitable cigarette tobacco, which involved the growth of a large number of kinds, both American and Indian, has been successful. An indigenous type—known as Pusa 28—has been found which fulfils all the conditions. This tobacco is a rapid and robust grower and gives a high yield of leaf of good colour, texture, and flavour, when cured in the country fashion with the smallest possible quantity of moisture. It s remarkable in its power of adaptation to widely different conditions and has done well not only in Bihar, but also in Burma, the Central Provinces, Central India, and the United Provinces. Seed for over 250,000 acres has already been distributed. In this work, the advantages of a supply of well-grown and carefully stored tobacco seed, in producing strong, uniform plants in the nursery and an even stand in the field, were instantly perceived by the growers, and no difficulties have been experienced in getting this type taken up on the large scale. During the search for a cigarette tobacco, one significant result was obtained. Without exception, all the American and other exotic varieties tried proved to be much less robust and grew much more slowly at all stages of growth than the indigenous types. Time was lost in the nursery; the plants had to be set out late and always failed to make full use of the growing season. This is fatal in tobacco-growing in India. There is only just time for a full crop to mature and for the curing to be completed, so that any tendency to slow growth rules a variety, however promising otherwise, completely out of court.

In the Rungpur District of Bengal, a belt of open soil lying over a deep sand layer occurs, on which tobacco is an important crop and where the fillers of the Burma cheroots of commerce are grown. In this area, a tobacco farm was started some years ago at Burirhat, near Rungpur, for the purpose of developing a local industry for the manufacture of cigars. The wrapper leaves are obtained from the Sumatra variety, the fillers from the local types. Cigars of good quality have been manufactured for some time, and efforts are now being made to induce the people to take up the industry. At Guntur in the Kistna delta of Madras and in Gujerat, tobacco farms have recently been started, but these have not been in operation long enough for the discovery of definite improvements.

## BIBLIOGRAPHY

Administration report of the Department of Agriculture in Bengal, 1921–2, Calcutta, 1923.

Howard, A.—The manurial value of potsherds, *Agr. Jour. of India*, xi, 1916, p. 256.

Howard, A., and Howard, G. L. C.—Studies in Indian Tobaccos. No. 1. The Types of *Nicotiana rustica*, L., yellow flowered tobacco, *Mem. of the Dept. of Agr. in India* (*Botanical Series*), iii, 1910, p. 1.

————————Studies in Indian Tobaccos, No. 2. The types of *Nicotiana Tabacum* L., *Mem. of the Dept. of Agr. in India* (*Botanical Series*), iii, 1910, p. 59.

————————The improvement of tobacco cultivation in Bihar, *Bull. 50, Agr. Research Institute, Pusa*, 1915.

Howard, G. L. C.—Studies in Indian Tobaccos, No. 3. The inheritance of characters in *Nicotiana Tabacum* L., *Mem. of the Dept. of Agr. in India* (*Botanical Series*), vi, 1913, p. 25.

————————Studies in Indian tobaccos, No. 5. The inheritance of characters in *Nicotiana rustica* L., *Mem. of the Dept. of Agr. in India* (*Botanical Series*), xiii, 1924.

Howard, G. L. C., and Kashi Ram.—Studies in Indian tobaccos, No. 4. Parthenocarpy and parthenogenesis in two varieties of *Nicotiana Tabacum* L., *Mem. of the Dept. of Agr. in India* (*Botanical Series*), xiii, 1924.

Proceedings of the Board of Agriculture in India, Calcutta, 1906.

Shaw, F. J. F.—*Orobanche* as a parasite in Bihar, *Mem. of the Dept. of Agr. in India* (*Botanical Series*), ix, 1917, p. 107.

Watt, G.—Dictionary of Economic Products of India, v, 1891, p. 302.

————The Commercial Products of India, London, 1908, p. 793.

# XVII

## PULSES

THE cultivators of India have been provided by Nature with a great variety of pulses. The country affords a remarkable example of the adaptation of the leguminous crop to a wide range of conditions as regards climate, soil, and water-supply. Legumes are met with on every type of soil, including those of the rice areas, from the Himalayas to the southern extremity of the Peninsula. These crops are a necessity in the rural economy of India, because they supply the proteids which are essential in the diet of the cultivator and of his oxen. The people consume a good deal of the seed. Some of the remainder and all the dried stems, leaves, pods, and seed-coats fall to the work cattle and to the buffaloes. The surplus seed is exported. Besides providing an essential portion of the food of both man and beast, leguminous plants play an important rôle in helping to keep up the fertility of the soil. All these crops possess root-nodules, which fix varying amounts of atmospheric nitrogen.

Notwithstanding their importance and the variety of interesting problems they offer for investigation, the pulse crops of India have not been studied in any great detail. Perhaps one reason for this neglect is the influence which the work of Hellriegel, on the part played by the root-nodules, has exercised on the teaching of agricultural science. Like all great advances in knowledge, the elucidation of the rôle of the nodule in nitrogen metabolism undoubtedly placed the succeeding generation of investigators in bondage, and materially restricted their independence of vision. Because the organisms in the nodule undoubtedly fix nitrogen and so enable leguminous crops, *grown under laboratory conditions*, to form flowers and seeds without any other supply of this essential element, the view has been tacitly accepted that cultivated leguminous crops can be lumped together as fixers of nitrogen and treated as agents of soil improvement which need no nitrogenous manure.

The pulses have undoubtedly been set apart as special crops which play a positive part in the maintenance of soil fertility. Recent researches, however, have drawn attention to the important fact that leguminous crops vary very greatly in their need of combined nitrogen, and cannot any longer be grouped together. It is only when the leguminous crop is grown for green-manure that the soil undoubtedly gains in nitrogen. In other cases, there is only a gain if the amount of nitrogen fixed by the nodules is greater than that removed in the crop. It is not always realized that this difference may be a minus quantity and that the soil actually loses in nitrogen by the growth of the leguminous crop. Two examples will illustrate the gain or loss in nitrogen which may be involved. Gram (*Cicer arietinum* L.) does best on poor land when forced to depend on the nodules for nitrogen. This leguminous crop improves the soil and complies fully with the teaching of the text-books. At the other extreme is Java indigo, which seriously depletes the supply of combined nitrogen and behaves very much like a cereal. The soil loses more nitrogen than it gains. These and similar results show that each leguminous crop must be studied separately. They also explain why nitrogenous manures are so frequently used in India for leguminous crops. The amount and kind of organic matter best suited to stimulate growth without interfering with fixation offers a fruitful field of study.

A perusal of the Indian literature, combined with practical experience in their growth, suggests that, besides gram, at least three other leguminous crops, namely, the pigeon pea (*Cajanus indicus* Spreng.), *mung* (*Phaseolus radiatus* L.), and *urd* (*Phaseolus Mungo* L. var. *Roxburghii* Prain), have a definitely beneficial effect on the fertility of the soil. The pigeon pea (*rahar*) is perhaps the most important cold-weather pulse of the alluvium and of the northern areas of Peninsular India. *Mung* is widely grown in the rains on the black soils, while *urd*, another monsoon crop, is one of the most esteemed of all the pulses of India, and can be grown before sugar-cane without affecting the yield of sugar. Both *mung* and the pigeon pea are deep-rooted plants, and, therefore, introduce an additional factor besides nitrogen fixation. How far is the undoubted benefit derived from the growth of these two crops due to nitrogen fixation and how far to the effect of the deep root-system on the texture and content of organic matter of the deeper soil layers? These matters, as well as the part each pulse plays in the nitrogen balance sheet, are questions which most urgently call for investigation. Detailed studies of the root-systems of the pulse-crops, in relation to the various soil types, appear essential for the better understanding of the part played by these plants in the maintenance of fertility.

Another aspect of Indian pulses, which would well repay investigation, is the improvement of the fodder supply. Most, if not all, of the pulses are already used for fodder, but practically nothing has been done in the intensive cultivation of these plants. Investigations are in progress, at Pusa and elsewhere, to ascertain which is the most profitable species to grow. A botanical survey of the most promising species should then be undertaken, with a view to the isolation of the best unit species for intensive cultivation. Such a study should include an investigation of the nitrogen cycle and the degree of response of the various unit species to intensive cultivation.

Before the pulses of India can be made to yield the maximum seed crop, a great deal of investigation will have to be undertaken. The methods of pollination and the occurrence of natural crossing in the field have only been worked out in a few cases. Such studies are particularly necessary in an order in which so many different methods of pollination are possible, ranging from cases like sann-hemp (*Crotalaria juncea* L.), where the visits of insects are necessary to produce setting, to close pollination in which natural crossing is exceedingly rare as in the case of gram. A further complication is introduced by the influence of humidity on the setting of seed and the consequent necessity of a long flowering period to ensure a full crop. This is one of the reasons why pure lines of the pigeon pea, under Pusa conditions, have always given a much lower yield than a mixture of types with a longer flowering period. Fortunately, the improvement of these crops does not involve questions of grain quality beyond the colour of the seed. In peas and gram, for example, white or yellow seeds are preferred to darker tints. Beyond such simple matters, the plant breeder need not, as a rule, concern himself with anything beyond an increase in the total weight of produce.

In gram, which is self-fertilized, mixed cultures are not required. It has been possible in this crop to isolate pure lines, which give far higher yields than the country crop. Several of these improved kinds are now being distributed to cultivators, for example, Malidah in the Central Provinces, and Pusa 17, 18, and 25 in the United Provinces. In Burma, where the gram is severely damaged by wilt, a selection from a sample of gram obtained from Karachi is proving to be immune to wilt. This variety is being rapidly taken up by the people.

As a large proportion of the pulses shipped to Europe are fed to stock, it is important to keep in mind the possible occurrence of cyanogenetic glucosides and other poisonous substances in any seeds exported. At least two of these crops are known to contain harmful substances. In the case of the beans (*Phaseolus lunatus* L.) grown on the dry, upland areas of Burma during the cold weather, and formerly exported under the name of Burma or Rangoon beans, the seeds contain cyanogenetic glucosides in varying amounts. Cases of stock poisoning have occurred in Europe from the use of these seeds. The demand has fallen off, and Burma beans now command relatively low prices. This difficulty is under investigation by the Burma Agricultural Department, and a number of interesting results have been obtained. The various unit species of *Phaseolus lunatus* have been found to differ in their total content of hydrocyanic acid; the development of the glucoside seems to depend on soil conditions, while the production of the poison is helped by the age of the seeds and by storage in a damp atmosphere. Work is now in progress in Burma in two directions—to discover a type of Burma bean containing the minimum amount of poisonous substance and also to ascertain whether some other pulse can take the place of *Phaseolus lunatus*. This case is of interest in showing the necessity of a constant study of the European markets, when developing the export trade in raw materials like seeds. Cases of poisoning by the use of these beans in Burma hardly ever occurred locally, as the seed was either boiled or parched before it was eaten. The heat prevents the formation of the poisonous principle. It was only when the beans were fed raw to stock, in the form of meal or cake, that casualties occurred, and the existence of cyanogenetic glucosides was discovered. After this experience, the possibility of the occurrence of such poisonous substances in all new pulses, grown for the export trade, will have to be borne in mind.

Besides Burma beans, another of the pulse crops of India—*khesari* (*Lathyrus sativus* L.)—produces in man a form of paralysis known as lathyrism. This crop often follows rice, and is grown in low-lying localities where other leguminous crops, such as the pigeon pea, will not thrive. It is only eaten by poor people ; in places where the cultivators are well-to-do the seed is fed to cattle. Lathyrism is common in most of the areas where *khesari* is grown. Once the people are affected, the partial paralysis continues and no cure seems to be possible. Recent investigations, carried out at Kasauli, Dehra Dun, and Pusa under the auspices of the Medical Research Fund Association, have shown conclusively that the seeds of *Lathyrus sativus* are quite harmless and contain no poisonous substance. The cause of the trouble has been traced to the seeds of a vetch—*Vicia sativa* L., var. *angustifolia*—which often occurs mixed with *khesari*. Hence in the future export of pulses from India, care must be taken, not only to see that the crop itself is free from poisonous substances, but also to pay attention to the various weeds with which it may be contaminated.

In the domain of vegetable pathology, one of these pulses—the pigeon pea—presents an interesting problem. All over the alluvium where this crop is grown, a number of plants frequently wither and die after they are full grown. This wilt disease has been investigated, and has been found to be caused by the inroads of a fungus which interferes with the upward transport of the sap to such an extent that the plants die for want of water. In some cases, as much as a quarter of the crop is lost ; in others, only a few plants here and there are affected. The relations between this disease and the soil conditions are now under investigation at Pusa, and are likely to throw a considerable amount of light on the influence of soil texture on the incidence of wilt. It will be interesting to know why certain plants in a field are destroyed while their neighbours escape, and why certain areas always favour the parasite.

## BIBLIOGRAPHY

Anderson, L. A. P., Howard, A., and Simonsen, J. L.—A preliminary note on lathyrism, *Indian Jour. of Med. Research*, x, 1923, p. 857.

Butler, E. J.—The wilt disease of pigeon pea, *Mem. of the Dept. of Agr. in India* (*Botanical Series*), ii, 1910.

Charlton, J.—The determination of prussic acid in Burma Beans (*Phaseolus lunatus*), *Bull. 140, Agr. Research Institute, Pusa*, 1922.

Duthie, J. F., and Fuller, J. B.—Field and garden crops of North-West Provinces and Oudh, Roorkee, 1882–3.

Ghosh, A. C.—The fodder pulses *meth*, *bhringi*, and *mashyumkalai*, *Bihar and Orissa Agr. Jour.*, v, 1917, p. 15.

Howard, A., and Howard, G. L. C.—Some aspects of the Indigo industry in Bihar, *Mem. of the Dept. of Agr. in India* (*Botanical Series*), xi, 1920, p. 1.

Howard, A., Howard, G. L. C., and Abdur Rahman Khan.—Some varieties of Indian gram (*Cicer arietinum* L.), *Mem. of the Dept. of Agr. in India* (*Botanical Series*), vii, 1915, p. 213.

———————————————Studies in the pollination of Indian crops, I, *Mem. of the Dept. of Agr. in India* (*Botanical Series*), x, 1919, p. 195.

McKerral, A.—*Fusarium* wilt of gram in Burma and measures taken to combat it, *Agr. Jour. of India*, xviii, 1923, p. 608.

Report on the Operations of the Burma Department of Agriculture for the years ending June 30th, 1922 and June 30th, 1923, Rangoon, 1923 and 1924.

Sil, S. N.—Improvement of rahar by selection, *Bihar and Orissa Agr. Jour.*, i, 1913, p. 25.

The nitrogen problem in Indian agriculture, *Proc. Tenth Indian Science Congress*, Asiatic Soc. of Bengal, Calcutta, 1924, pp. 243–261.

Thompstone, E., and Sawyer, A. M.—The peas and beans of Burma, *Bull. 12, Dept. of Agr., Burma*, 1914.

Watt, G.—The Commercial Products of India, London, 1908.

Warth, F. J.—A note on hydrocyanic acid in the Burma bean (*Phaseolus lunatus*), *Mem. of the Dept. of Agr. in India* (*Chemical Series*), vii, 1923, p. 1.

Warth, F. J., and Ko Ko Gyi.—Prussic acid in Burma beans, *Bull. 79, Agr. Research Institute, Pusa*, 1918.

## XVIII

## FRUIT-GROWING

THE development of fruit-growing in India, on Western lines, is one of the subjects which periodically come up for discussion in this country. It is a popular topic, and one which possesses a certain fascination of its own. Several important fruit-growing areas exist in India, while others are being developed. Saharanpur and Amritsar are well-known centres of fruit cultivation. In parts of Tirhoot, plantations of mangoes and litchis are numerous, and in favourable years the export of fruit is considerable, in spite of the fact that there is a break of gauge on the railway to Calcutta. Another large mango industry exists in the Bombay Presidency. The cultivation of bananas in parts of the Konkan and in the neighbourhood of Calcutta is considerable, while in the Central Provinces the orange industry shows signs of expansion in the neighbourhood of Tharsa. Near Allahabad, guavas are largely grown for general distribution. In the hill regions, temperate fruits such as apples, pears, peaches, and grapes have been established at a number of centres, such as Kulu, Almora, Kashmir, the North-West Frontier, and Baluchistan.

In several respects, fruit-growing in India labours under great disadvantages. Transport is expensive, cold storage has not emerged from the stage of discussion, while the various methods of preserving the surplus fruit are still in their infancy. Except in large cities like Calcutta and Bombay, purchasers willing to pay high prices for choice fruit do not yet exist in large numbers. Thus the chief factors which have led to the modern developments of intensive fruit culture in the West—namely, cheap, rapid, and reliable transport to great centres of population, cold-storage facilities at the markets, a multitude of well-to-do customers, as well as methods of preserving the surplus crop (such as jam making, bottling, and evaporation)—remain to be developed in India. Mainly for these reasons, fruit-growing is still carried on on primitive lines, and little or no intensive cultivation of superior varieties has been developed. The country, however, possesses some advantages. The population is large and mostly vegetarian, while, in the cities, the standard of living is steadily rising. Provided suitable varieties are selected and properly grown, large crops are possible.

A great deal still remains to be done before any great advance in fruit-production in India can be achieved. Obviously the first condition of progress, in dealing with an ancient agriculture like that of India, is to gather together the existing knowledge and experience and to set this out in concrete form in its relation to the main growth factors. This is always necessary in India, where there is little or no indigenous literature on such questions, and where knowledge is handed down in the various localities mainly by word of mouth. A survey of the varieties of the various kinds of fruit grown and their relation to soil and climatic conditions, of the methods of propagation used by the people as well as of the details of cultivation, is required. Excellent varieties exist, and a good deal of experience and knowledge of the various aspects of fruit-growing is to be met with in many localities, particularly in the various Horticultural Gardens. As it has not been recorded in a readily accessible form, it is not available for the use of the grower or the investigator. A useful beginning, however, has been made in the case of the mango by Burns and Prayag, but a great deal more work of this character remains to be done to clear up the present position and to make a definite starting-point for further work.

While little is known of indigenous fruit-growing, still less has been ascertained on the important question of the relation of the root-development of the various species of fruit trees to the soil type, water-supply, and soil temperature. A few investigations on this subject have been published, but little more than the fringe of the subject has been touched. A fruit plantation is a permanent crop. Once it is established it cannot be altered except at great cost. In fruit-growing, it is even more important than in the case of annual crops to see that the roots and the soil are in suitable relation. If they are not, it is practically impossible, with the means at the disposal of the grower, to make up for such a fundamental disadvantage. The subject of the root-systems of the various fruit trees needs to be investigated from the seedling stage right up to the time the trees are in full bearing. The distribution of the roots, as well as the seat of root activity throughout the year, should be ascertained. This information can only be determined when we know where and when the root-hairs are active. Such an investigation includes the root-development of the various stocks on which fruit trees are propagated and also the effect of the various types of scion on this development.

A good deal of work has been done in discovering the best stocks. In the Peshawar valley, the stock problem has been investigated in the case of the orange when grown on alluvial soils with abundant irrigation water. The rough lemon was found to give the best results with the Malta orange, while the sweet lime was the most suitable for the *sangtara*. In Baluchistan, it was found in the case of the peach, plum, cherry, and apricot that practically all the introduced varieties had been worked on stocks quite unsuitable for the high soil temperatures which obtain in the Quetta valley. Far better results were obtained by the use of stocks like the almond; mahaleb, mariana, and myrobolan.

The nature of the union between stock and scion influences both root and shoot. The operations of grafting and budding act on the combination like a severe pruning of both the shoot and the root. The junction is always more or less imperfect, and interferes with the supply of crude sap to the leaves, in one direction, and in the return of elaborated food material to the roots in the reverse direction. The first leads to restricted vegetative growth, and so automatically prunes the tree. The interference with the return of elaborated food from the leaves to the roots naturally brings about root-pruning. As the formation of wood and fruit are antagonistic, the operation of budding or grafting is, within limits, bound to tend towards fruit-production. Sometimes, as in the case of the orange, the interference caused by the union is too great and both root and shoot development suffer, with the result that either the development of the shoot or of the root or of both is too restricted for efficient growth. All these questions need investigation.

Information is required as to which combinations are effective, which are useless, and the precise reasons for these results. There is a very practical side to these inquiries, namely, the reasons why so many budded plants, even when the stock and scion have been properly selected, do not thrive after planting out. From preliminary results obtained at Quetta with the peach budded on the almond, it was found that there is every gradation possible between stunted, diseased plants which never bear and exceedingly vigorous trees. It was often observed that the behaviour of the plant in the nursery persisted throughout its after life and that no recovery ever took place. This was found to be associated with the width of the callus between the stock and the scion. If the width of callus was small, the budded tree throve well ; as it lengthened, the vigour of the combination fell off. The results pointed to the need for the greatest care in nursery work generally and to the most rigorous selection of plants for setting out. When the nursery work in India has reached the proper standard, a further advance is possible (as the recent American results on the orange prove) by the careful selection of the mother tree and the use of the most suitable bud-wood from such trees.

Another matter arising out of the study of root-development is the effect of heavy soils on the growth of the orange in Peninsular India. The general experience is that the well-being of an orange plantation on the black soils depends largely on the permeability of the subsoil, and unless this is provided, the trees die and the plantation becomes unthrifty. Is this due to the fact that the lower roots are active after the rains and that on deep black soils the aeration at this period is too poor for active growth ?

A subject closely related to root-development is the influence of high soil temperatures on the active rootlets. The effect of high temperature on the leaves and shoots of plants is automatically regulated by transpiration. The leaf is able to keep itself cool. Very different, however, is the case of the active root. It may be that during the hot weather, when the direct rays of the sun heat the soil, the active roots are unable to withstand the high temperature unless the soil is either kept cool by irrigation water or by shading. So far very little has been done on this question in India. In only one case, namely, the cultivation of apricots in the Quetta valley, does any attention seem to have been paid to this matter. Here it was found that apricot seedlings invariably withered during the period of high temperatures of June and July unless the soil was kept cool by means of a deep vegetable mulch. The people round Quetta invariably establish an apricot orchard by planting the seedlings among lucerne, a crop which is frequently irrigated. The trees are also planted very close, so that they soon form a complete canopy and shade the soil. This practice of close planting is so universal in India and seems so deeply rooted in the instincts of the people that it is more than probable that it is based on some fundamental growth factor. This may be the shading of the soil so as to keep the temperature within proper limits.

Another general question connected with fruit-culture is the effect of grass on the growth of the trees. This problem has been under experiment at Pusa for a number of years, and very striking results have been obtained. All the fruit trees grown, namely, plums, custard apples, limes, loquats, mangoes, litchis, guavas, and peaches, were adversely affected by grass very much like the apple at Woburn in Great Britain. Some of these species, namely, the plum, custard apple, peach, lime, and loquat, die out altogether under grass. Litchis and mangoes are stunted, and only maintain themselves with the utmost difficulty. The guava is the least affected. The various factors involved are being studied, including the question of the formation of a soil-toxin by the grass.

After the selection of varieties and the cultivation of the trees, the most important matter for investigation in India is the transport of fruit. At present, the Indian cultivator often solves the difficulties of packing and transporting fruit by the simple device of picking it unripe and allowing it to ripen afterwards. By this means, losses by theft, birds, and animals in the gardens are avoided, and careful packing is unnecessary. Nevertheless, the losses are considerable, and the flavour and appearance of the product leave much to be desired. Even when the fruit is left to ripen on the tree and modern methods of packing are adopted, a new set of difficulties make their appearance, namely, thefts in transit on the railways. This greatly interferes with the development of the fruit trade. To reduce these practices, the packages have to be very elaborate and have to be fastened and sealed so that the detection of theft is an easy matter. On the other hand, fruit will not travel well except in ventilated packages. To check petty pilfering through the interstices these have to be kept small, and so a great deal of extra wood has to be used in the construction of the packages. All these precautions involve the expenditure of time and money, and add seriously to the cost to the consumer. A good deal of attention has been paid to the improvement of fruit packing in Baluchistan. Various packages for small and large consignments of peaches, apricots, plums, tomatoes, and grapes have been designed, tested, and introduced into general use. Two transport concessions have been obtained for the fruit trade—the grouping of packages to the same consignee for purposes of charge and the free return of empties of standard pattern from all railway stations in India to Quetta and Chaman. These improvements have had a favourable influence on the fruit trade generally, and especially on grape cultivation in Baluchistan. Formerly, the grape crop could only be transported to places near the Frontier. When sent in the new crates they travel without damage to all parts of India, a fact which has increased the demand and therefore the prices obtained locally by the growers. The principles underlying the construction of these crates have been widely adopted, and copies in indigenous materials are now to be found in all the great markets of India. Perhaps the best way of getting the most out of these improved packages, and at the same time checking petty pilfering, will be to develop the present Frontier trade so that sealed vans can be dispatched to large cities like Calcutta and Bombay.

After transport, questions of storage and preserving follow in order of importance. Except in a few hotels, there are no cold-storage facilities in India, and, so far, refrigerator cars on the railways have not proved a success under local conditions. In America and Europe, cold-storage has been developed chiefly for such articles of food as meat, fish, and hard fruits like apples and pears. Before cold-storage and refrigerator cars can be adopted for Indian fruit, a great deal of work must first of all be done with the indigenous varieties, and suitable methods of cooling worked out on the spot. It is obvious that methods suitable for apples will have to be considerably modified before they can succeed with more delicate fruit, such as mangoes, grapes, bananas, and peaches, which easily lose their flavour. Moreover, the life of Indian fruit in cold-storage might be much less than is the rule with apples.

Very little work has so far been done on the preserving of Indian fruits in seasons when the markets are glutted. Apricots and grapes are sun-dried in a somewhat primitive fashion on the Frontier. A few small factories exist for the tinning of mangoes and litchis, and in the hills a certain amount of jam is made by private individuals. The great difficulties to be overcome in this work are the high price of the materials used (such as tin-plate, bottles, solder, and sugar), the rough character of the labour available, and the competition of the uniform and attractive products of Australia and California, which found their way into the Indian markets during the war.

## BIBLIOGRAPHY


Amin Khan, M.—Ring budding, *Agr. Jour. of India*, viii, 1913, p. 74.

Basu, B. S.—Orange cultivation in the Khasi Hills, *Agr. Jour. of India*, i, 1906, p. 62.

Brown, W. R.—Peach growing around Peshawar, *Agr. Jour. of India*, vi, 1911, p. 14.

———Grape growing around Peshawar, *Agr. Jour. of India*, vi, 1911, p. 221.

———A trial of orange stocks at Peshawar, *Agr. Jour. of India*, ix, 1914, p. 84.

———The orange: a trial of stocks at Peshawar, *Bull. 93, Agr. Research Institute, Pusa*, 1920.

Burns, W.—The Pairi mango, *Agr. Jour. of India*, vi, 1911, p. 27.

———The strawberry culture of Mahabaleshvar, *Agr. Jour. of India*, ix, 1914, p. 288.

Burns, W., and Prayag, S. H.—The classification of mango varieties, *Agr. Jour. of India*, x, 1915, p. 374.

———The book of the mango, *Bull. 103, Dept. of Agr., Bombay*, 1921.

Burns, W., and Kulkarni, L. B.—Some observations on the roots of fruit trees, *Agr. Jour. of India*, xv, 1920, p. 620.

Burns, W., and Joshi, P. G.—The drying of bananas, *Agr. Jour. of India*, xv, 1920, p. 166.

Cheema, G. S., and Gandhi, S. R.—The influence of notching on the yield of fig trees, *Agr. Jour. of India*, xviii, 1923, p. 501.

Cleghorn, J.—Melon culture in Pishin, Baluchistan, *Agr. Jour. of India*, ix, 1914, p. 124.

Haller, G.—Orange cultivation in Coorg, *Agr. Jour. of India*, i, 1906, p. 127.

Hartless, A. C.—Mango crops and some factors influencing them, *Agr. Jour. of India*, ix, 1914, p. 141.

Howard, A., and Howard, G. L. C.—First and second reports of the Fruit Experiments at Pusa, *Bulls. 4 and 16, Agr. Research Institute, Pusa*, 1907 and 1910.

————————Some improvements in the packing and transport of fruit in India, *Bull. 2, Fruit Expt. Station, Quetta*, 1919, and *Agr. Jour. of India*, viii, 1913, p. 274.

————————The cultivation and transport of tomatoes in India, *Bull. 1, Fruit Expt. Station, Quetta*, 1913, and *Agr. Jour. of India*, viii, 1913, p. 245.

————————The improvement of fruit culture in Baluchistan, *Bull. 9, Fruit Expt. Station, Quetta*, 1919.

Joshi, R. S.—Orange cultivation in the Central Provinces, *Agr. Jour. of India*, ii, 1907, p. 64.

Kulkarni, L. B.—The drying of plantains at Agashi, *Agr. Jour. of India*, vi, 1911, p. 289.

————Investigations on papaya, *Agr. Jour. of India*, ix, 1914, p. 384.

Narayan Rao, D. L.—Mango cultivation, *Agr. Jour. of India*, vi, 1911, p. 405.

Prayag, S. H.—The alphonse mango, *Agr. Jour. of India*, viii, 1913, p. 169.

————Mango culture in Goa, *Agr. Jour. of India*, viii, 1913, p. 172.

Rout, E. L.—How to bottle fruits, vegetables, &c., for domestic and commercial purposes, *Agr. Jour. of India*, xi, 1916, p. 384.

The Duke of Bedford and Pickering, S. U.—Science and fruit growing, London, 1919.

## XIX

# DISEASE IN PLANTS

DURING the last twenty years, a great deal of work has been done in India on the diseases of crops in which insects or fungi are involved, and a large number of valuable papers have been published on the subject. The life-histories of the parasites have been investigated in detail, the nature and amount of the damage done have been described, and various remedial measures have been suggested. In some cases, schemes for the control of pests have been elaborated and put into practice. In addition to the diseases in which insects or fungi are associated, a number of others, in which parasites do not appear to be involved, such as the wilt of indigo in Bihar and the spike disease of sandalwood in Mysore, have been explored. It is not proposed to describe these various researches in detail, but to deal with some of the more general aspects of disease.

Broadly speaking, two lines of attack are possible in dealing with the diseases of crops: (1) direct methods, which include the destruction of the parasite and the prevention of infection, and (2) indirect methods, such as the avoidance of disease altogether by the selection of disease-resistant varieties and by the increase of the natural resistance of the plant. It will be interesting, in the light of Indian conditions and of the present state of our knowledge of disease, to discuss these two methods separately.

The destruction of the parasite is simplified when the facts of its life-history are known. It is generally much more vulnerable at one point than it is at others. We can, for example, often cut short the activities of a fungus by spraying the crop with some wash containing copper salts which destroy the germinating spores. Seedlings can be protected by treating the seed, before it is sown, with copper sulphate. In the case of pests like green-fly, we can suffocate the insect by spraying the infected plants with an emulsion containing soft soap. In other cases, the insect or fungus can be collected in the resting stage and burnt. When, however, we come to the application of this group of remedies to Indian conditions, a number of difficulties arise. The ignorance of the people makes it impossible for them to realize the meaning of plant diseases and to grasp the reasons for the treatment advocated. Their poverty is a great bar to the adoption of such costly devices as spraying machines and sprays as a part of their everyday practice. The greatest obstacle to the adoption of methods aimed at destroying the parasite is the comparatively small value of the crop to be saved. To the cultivator the game does not seem worth the candle. He prefers to accept the visitation of pests as part of the general order of things rather than attempt costly and uncertain remedial methods. His ignorance of the real nature of disease, his desire to obtain the last ounce out of his small holding make methods based on the destruction of affected plants unpopular. There are of course exceptions. In the case of permanent crops like tea, coffee, rubber, coco-nuts, and fruit, a moderate amount of capital is often available for the treatment of diseases, and the consideration of remedies comes into practical politics. Such crops are so valuable that the benefits likely to be gained can be balanced against the cost of the treatment. The greatest scope for this class of remedies, however, is met with in really valuable crops like the hop, where the capital involved is very great and where it is all-important to keep the ripening flowers free from green-fly and mildew. Systematic spraying and powdering is the rule in hop-growing, and the results obtained more than justify the cost.

The second line of attack in dealing with diseases is prevention. This can be achieved either by changing the variety or by improving the agricultural conditions, so that the vigour of the plant is increased to such an extent that it is able to overcome the parasite. In the case of the sugar-cane in Java, these methods are now solely relied upon in combating disease. The Indian Sugar Committee, when dealing with this aspect of the Java sugar industry, stated: 'It is held that proper methods of cultivation and the introduction of good varieties are the most important factors in the control and elimination of diseases.' This passage is significant. The Java sugar industry corresponds to hop-growing in Great Britain. It is perhaps the high-water mark of tropical agriculture, and owes its present position to various natural advantages, which have been developed to the utmost by the energy and initiative of the Dutch planters, supplemented by the efforts of a succession of highly qualified scientific investigators, who have explored, in the greatest detail, the various directions in which the production of sugar can be increased. In the early days of this development, considerable attention was paid to the insect and fungoid diseases of the cane. It is most significant that direct methods have been given up, and that attention is now concentrated solely on the variety and on its proper cultivation. A recent visitor to Java, on his return to India, stated that an outbreak of the red-rot fungus on a sugar estate in that colony would involve the dismissal of the manager, because experience has shown that this disease only occurs as a result either of improper methods of agriculture or of the cultivation of unsuitable varieties.

There seems little doubt that the most promising method of dealing with diseases of crops in India is by means of the plant. In this country, as in the case of the sugar-cane in Java, the varieties cultivated differ enormously in resistance to disease. Thus in the early period of the sugar industry in North Bihar, all kinds of canes were tried by the planters, but most of these have had to be given up owing to the attacks of the red-rot fungus. At first an indigenous cane, known as *Burli* or *Hemja*, was found to resist the fungus sufficiently well to enable the industry to take root. *Burli* is now being replaced by the new seedlings derived from crossing canes with *Saccharum spontaneum*. Here the best remedy in dealing with disease has been to change the variety rather than to destroy the fungus or protect the plant against its attacks.

Resistance to disease can be increased or decreased by change of soil and by alterations in soil texture. A few examples will make this clear. The red-rot of sugar-cane is very common in certain tracts of India, such as the black soil areas of the Central Provinces, the alluvium of North Bihar, and the Godavery delta in Madras, in all of which the aeration of the soil is poor at certain periods of the year. This fungus attacks the cane during the ripening period, and leads to a great loss of sugar. Now when the varieties of sugar-cane grown on the heavy black soils at Sindewahi in the Central Provinces were transferred to the permeable, well-aerated *bhata* soils at Chandkhuri in the Chhattisgarh Division with the same rainfall, two remarkable facts were observed. In the first place, the sugar-cane grew much faster on the poorer *bhata* than on the much richer black soil, and yields of as much as forty tons of stripped cane to the acre were obtained. In the second place, and this fact is of the greatest interest, there was a remarkable absence of the red-rot fungus. As is well known, the texture of heavy black soils, like those at Sindewahi, is easily destroyed during the heavy monsoon, percolation is reduced, aeration suffers, and the growth slows down. In the case of the sugar-cane, this was followed by an attack of red-rot. At Chandkhuri, on the other hand, a rainfall of 70 inches has no deleterious effect on the texture of *bhata*. Growth is not affected, and there is no disease. This experience agrees in all respects with that of Java. An attack of red-rot follows the destruction of the tilth, and thus appears to be one of the consequences of poor soil-aeration. Obviously a simple line of attack is to preserve the tilth where possible rather than attempt the destruction of the fungus.

It is possible to increase disease-resistance in the same locality by altering the treatment of the soil. Thus at Quetta, in the case of the almond and the peach, heavy winter irrigation was found to render these plants exceedingly susceptible to the attacks of green-fly, while normally grown trees side by side were practically immune. Deep cultivation and thorough soil-aeration, after the green-fly attack developed, were followed by greatly increased resistance to the disease. The first-formed leaves on these trees showed extensive damage by the pest, the late-formed foliage on the same twig was normal and perfectly healthy. The disease never spread from the old to the new leaves.

A collection of varieties from various tracts often contains kinds which differ greatly in immunity to disease. These differences are associated with the kind of gearing between roots and soil. An interesting example occurred at Pusa in

1921, when certain deep-rooted varieties of a pulse known locally as *khesari* (*Lathyrus sativus* L.) were attacked by green-fly. In that year, deep-rooted varieties from the black soils, shallow-rooted types from the alluvium, as well as kinds with an intermediate root-system from the Allahabad District, were sown in the same plot. All the types from the black soil area were badly attacked by green-fly, all those from the alluvium were immune, while the types with intermediate root-development were only slightly affected. Although the infected and disease-free plots were often side by side, in no case did the *Aphides* leave the deep-rooted cultures and attack those with shallow roots. Really healthy plants seem able to keep insects and fungi at bay, and their juices seem unsuitable for the nourishment of these organisms. The influence of soil conditions on disease-resistance is seen when the wheats of the Peninsula are grown in the plains. Many of the deep-rooted varieties from the black soils, when grown at Pusa, are practically destroyed by rust. In their own locality, however, rust attacks are rare. Alterations in the environment, therefore, influence the natural resistance of a variety. On the other hand, when the shallow-rooted variety, known as Pusa 6, is grown at Pusa next to these black soil types, it always remains free from rust although surrounded by a mass of infection. A very interesting case of the natural antipathy between the active growth of roots and fungi is described by Barker in the 1920 report of the Long Ashton Research Station. When grown in water culture, the roots of certain fruit trees soon became covered with white fungoid growths. When, however, either lenticels or new roots or both were developed, the fungus disappeared and did not reappear till the end of the growing season when root-activity began to diminish. Active root growth and the presence of the fungus were found to be incompatible.

These examples, of which many others could be quoted, illustrate three facts, namely, (1) varieties differ greatly among themselves in resistance to disease, (2) alterations in soil treatment profoundly influence the natural resistance of the plant, and (3) when resistant and susceptible varieties are grown side by side the disease does not spread to the resistant varieties.

Recent investigations on this question have also established two other principles, namely, (4) resistance to various parasites can be inherited, and (5) the immunity to rust attacks of even the most resistant kinds of wheat, such as einkorn, breaks down when the growing conditions are sufficiently adverse.

What is the precise nature of disease-resistance ? Is it something inherent in the variety, is it one of the consequences of a happy relation between the plant and its environment, or are there two types of disease-resistance ? These questions are fundamental, and call for the most careful and detailed investigation. India is a favourable country for such work, as it possesses on the black soils and on the alluvium two quite different classes of varieties and two absolutely different soil types. The varieties of a crop like linseed from the black soils are deep-rooted, those from the alluvium are shallow-rooted. When the deep-rooted types of the black soils are grown on the alluvium their resistance to disease falls off. Is this solely because of the unsuitability of the gearing between plant and soil ? The facts relating to root-development appear to be important in the study of disease-resistance, and we require to know how far immunity and susceptibility are bound up with the fitness or the unfitness of the root-system for the particular soil type. Among other things it may be that resistance to disease is partly a consequence of efficient gearing and that the attacks of parasites only follow mistakes in the choice of varieties or in agricultural treatment. The experience of the sugar planters in Java certainly falls in with this view, which, if established by more examples, will lead to parasites being regarded not as pests to be destroyed but as very valuable indicators, provided by Nature, for checking the proceedings of the agriculturist. The appearance of a pest in the agriculture of the future may merely show (1) that the wrong variety is being grown, (2) that the crop is being grown in the wrong way, or (3) that Nature did not intend it to be grown at all.

The study of the diseases of crops is entering on a new phase. The investigation of separate diseases is slowly becoming merged into research on disease. In future, the parasites involved will tend to become little more than agents in studies which will include, besides disease, the general physiology of the plant, the life-history of the root-system, the incidence of the various soil factors, temperature, and humidity on the growth of the crop. The plant pathology of the future is likely to be conducted on a broad basis, and will thus be the means of integrating and bringing into proper focus much knowledge relating to crop-production which is at present fragmentary.

## BIBLIOGRAPHY

Barker, B. T. P.—Root development in newly planted trees. Annual report of the Long Ashton Research Station, 1920, p. 50.

Charles Ogilvie Farquharson and his work in West Africa.—*Kew Bulletin*, 1918, p. 353.

Clouston, D., and Padmanabha Aiyer, H. R.—The physical texture of soils in its relation to crop-production, *Agr. Jour. of India* (*Indian Science Congress Number*), xiii, 1918, p. 89.

Howard, A.—The influence of soil factors on disease resistance, *Ann. of Applied Biology*, vii, 1921, p. 373.

——————Disease in plants, *Agr. Jour. of India*, xvi, 1921, p. 626.

Howard, A., and Howard, G. L. C.—Note on immune wheats, *Jour. of Agr. Sc.*, ii, 1907, p. 270.

Report of the Indian Sugar Committee, Simla, 1921.

Sorauer, P.—Handbuch der Pflanzenkrankheiten, Bd. 1, Die nicht-parasitaren Krankheiten, Berlin, 1921.

Ward, H. M.—Disease in plants, London, 1901.

# PART III

# ORGANIZATION

THE benefits to be derived from the successful study of the crops of a country need no argument. The difficulty is to secure investigators with the requisite training, experience, and ability for the work. When these are found, it is necessary to provide them with the means and with such working conditions that they can devote the whole of their energies to research. The form of organization adopted must simultaneously fulfil two conditions: (1) freedom for the investigator, and (2) the proper control of the money provided for the work. Some compromise—in which the reasonable requirements of the men who do the work and the interests of the public, who provide the money, are reconciled—is needed if science is to take its proper place in the nation-building of the future. The difficulty is not so much the provision of funds as the happy union of ability and opportunity with the means required.

In the following chapters, some ideas as to the type of investigator needed for the problems underlying crop-production and on the organization of scientific work have been put forward. This has been done to stimulate discussion and thought rather than to suggest the ideal solution of these difficult questions.

## XX

## THE IDEAL INVESTIGATOR

In the preceding pages, the general nature of the problems of Indian agriculture has been indicated. These questions are complex and are quite unlike the ordinary subjects of academic research, where the factors and conditions can be accurately controlled. For their successful solution, they manifestly require every aid that a wide knowledge of science can give. In agricultural investigations, we can never get rid of the ever-changing environment. The prejudices of the cultivators, the smallness of the means at their disposal, and the general labour conditions impose a further set of limitations. In all directions, the investigator meets with well-defined working conditions. It is little wonder, therefore, that the improvement of Indian agriculture is such a difficult matter and that progress is so slow.

What kind of investigators are needed for such work, and what can be done in India to fit them for their task? It is now proposed to answer these two questions in so far as they relate to crop-problems.

What is the essential nature of a crop? It can be regarded as a group of living factories which utilize two classes of raw material: the one obtained from the soil, the other from the atmosphere. Various mineral salts, in dilute solution, enter the plant from the soil by way of the root-system and are carried to the green leaves by the upward transpiration current. From the atmosphere, oxygen and carbon dioxide reach the same point by way of the stomata. In the green cells of the leaf, these two classes of raw materials are worked up into complex food substances by means of energy focussed from the sun through the medium of the chlorophyll corpuscles. Unlike an animal, a plant has to make its own food before it can feed. In both cases, the actual food, however, is very similar. The crop is constantly manufacturing food, developing new organs, and completing its life cycle under varying conditions as regards the supply of raw materials, temperature, illumination, and humidity. The manufacture of its own food by the green leaves is the first work of the plant. Its second duty is to provide a surplus—in the shape of reserve materials which are often packed into the seed—for the use of the next generation. Man intercepts these reserve materials for his own use, and on their amount and quality the success or failure of crop-production depends. In this manufacture of food, it is well to bear in mind the fact that the plant has always to feed itself first of all and that the formation of reserves marks as it were a second stage of activity. Naturally, the greater the efficiency of the factory, the more food there will be for growth and the greater will be the volume of the reserves. The duty of the investigator of crop-problems is to study the working and output of this natural factory, to discover the directions in which it can be improved, and then to devise the most economical method of carrying this out. Often improvements are possible but they are not economic. The trouble and cost involved are prohibitive. Such discoveries, therefore, are only of academic interest and cannot affect practice. To be successful, the game must always be worth the candle. In India, the cultivators are mostly in debt and the holdings are small. Any capital required for developments has to be borrowed. A large number of possible improvements are barred by the fact that the extra return is not large enough to pay the high interest on the capital involved and also to yield a profit to the cultivator. It is this economic complication which makes it so difficult to improve production in India.

The knowledge required by the investigator of crop-problems must obviously be considerable. On the scientific side, he must be well trained in all branches of botanical science, including morphology, anatomy, physiology, pathology, systematic botany, ecology, and genetics. In addition, he must have a sound knowledge of general science, in which chemistry and physics should be included. Such knowledge is essential because, in crop-problems, it is not the plant alone that has to be studied but *the plant in relation to its environment*. This part of the training of the future investigator is the work of the University, and is best obtained in the ordinary science schools. One drawback of even the best scientific training has, however, to be borne in mind in the training of investigators, namely, the artificial sub-division of science rendered necessary by the exigencies of teaching and examinations. The advanced student should be made to realize that such sub-division, and the consequent existence of schools of botany, chemistry, physics, and so on, is after all an artificial thing which is not recognized by the living plant. The investigator of crop-problems should disabuse his mind of the compartment system necessary for the teaching of science, and not allow it to invade the second stage of his training—the experiment station phase. In agriculture, the work of the plant forms a definite branch of the subject and has no separate existence in terms of botany, chemistry, or physics. This is a most important matter because among scientific workers there is always the temptation for the disciples of each school to separate into something approaching a faction. Unfortunately, these academic conventions, which really belong to the teaching of science, have affected agricultural research and have tended to limit unduly the scope of much of the work which has been done on crops. This narrow outlook is, however, slowly breaking down, and crop-production, rather than one or other of the sub-divisions of agricultural science, is coming to be regarded as the real subject of study.

After an adequate scientific training, the future investigator of crop-problems must master the art of agriculture as far as it relates to crops, and also pay attention to a number of trade aspects. Here again, anything in the nature of separate compartments must be avoided. The art of agriculture, which is nothing more than the crystallized experience of generations of tillers of the soil, must be simultaneously looked at from the point of view of the cultivator and of the student of science. In ancient systems of agriculture, such as that of India, it must never be forgotten that the cultivator is generally sound in his procedure ; the difficulty often is to perceive the scientific basis of his practice. Similarly, in studying the trade aspects of crop-production, these must be welded into the scheme of things and not regarded as a separate subject. If the investigator is to use his science as an instrument for improving production, it is clear that he must consider all aspects of a crop like cotton from the moment the seed begins to germinate in the moist soil till the fibre is worked up in a factory into some material for the use of mankind.

Such is the training and experience needed on the part of the worker. It is obviously considerable. The University phase is the beginning and constitutes the period of acquisition, something like the caterpillar stage in the life-history of a butterfly. After this, practical experience is essential, and the future investigator must be at great pains to weld the two stages of his training into a well-balanced whole. Such a training, as has been indicated, must be expensive both as regards money and time, and any State organization which employs such men cannot possibly afford large numbers.

In India, experience has shown that two difficulties will have to be overcome in the training of investigators. The first occurs in the University phase, the second in the apprenticeship stage. Both these difficulties often arise from economic causes.

In the University or College period, many students show a marked tendency to rush into agricultural and other technical work before they have acquired a sufficiently advanced knowledge of the science they wish to make use of in later life. This leads to disappointment when the time comes to take up independent investigation. Sometimes an effort is made to repair the want of preliminary training by short periods of work in European and American Universities, but the youthful period of easy acquisition has been passed, and the results are frequently disappointing. The difficulty can only be overcome by continuing to raise the standard of science teaching, a process which has already begun in the Indian Universities, and by impressing on the students that the first requisite in the investigation of subjects like crop-production is adequate scientific knowledge, and that this can only be acquired during the student phase. It is too late afterwards.

The apprenticeship phase—between the University period and the actual work of research—is at present very defective. Sometimes this stage is omitted altogether and the University graduate takes up research work on all manner of applied subjects on leaving College. As a result, the subject suffers from the publication of obviously immature work. What is now required in matters concerning crop-production is the institution of a definite apprenticeship phase at the experiment stations. Here the young University graduate can study the art of agriculture, master the technical details of the work of investigation, and begin research in collaboration with men of experience. It is, however, very difficult, for economic reasons, for the average student to give the time required. This obstacle can be removed by the liberal provision of post-graduate studentships. A beginning has been made in this direction by the Central Cotton Committee. A number of research studentships have been founded to enable selected graduates of Indian universities to spend two years at approved experiment stations, where they can serve an apprenticeship in research work on cotton under investigators of experience.

## BIBLIOGRAPHY

Couchman, M. E.—The agricultural classes in Madras, and how to approach them, *Agr. Jour. of India*, iv, 1909, p. 43.

Howard, A.—The application of botanical science to agriculture, *Agr. Jour. of India (Indian Science Congress Number)*, xi, 1916, p. 14.

Howard, A., and Howard, G. L. C.—Post-graduate training in Agriculture, Pusa, 1922.

# XXI

## THE ORGANIZATION OF RESEARCH

AFTER the investigator of agricultural problems has been secured, working conditions must be arranged by which his maximum usefulness can be obtained. This involves the provision of funds and some form of organization. At the present time, there is general agreement that the development of the agricultural resources of a country is one of the duties of the State, and that the cost of the work should be met, in part at least, from public funds. The manner in which State assistance can best be given is, therefore, an important matter.

A number of experiments have been made, in many countries, in the encouragement of agricultural research by the State. These experiments fall into two main divisions: (1) assistance to existing institutions; and (2) the creation of official departments of agriculture. Examples of both these systems are to be found in the British Empire.

In Great Britain, State assistance to agricultural work takes the form of grants to a number of local autonomous institutions. Some, like the Rothamsted Experiment Station, were originally founded and endowed by private benefactors. Others, such as those at Cambridge, Long Ashton, Aberystwyth, and the Rowett Institute of Animal Nutrition at Aberdeen, owe their origin to the influence of Universities or County Councils and similar local bodies. The State, however, has assisted these centres of research—at first by grants from the Ministry of Agriculture and latterly from the Development Fund. As regards financial support, therefore, these Research Institutes possess two sources of income: (1) endowments or local grants; and (2) contributions from Imperial funds. All these institutions are governed in very much the same way—by Boards of Governors, Trustees, or Committees, whose duties consist in a general supervision of expenditure and in the appointment of the staff. There is little or no attempt at interference with the workers. It is generally agreed that the essential act of organization is the selection of the individual to do the work, and that after this, nothing useful is possible in the way of direction. The State is content with the very minimum of inspection. Carefully selected officials of the Ministry of Agriculture, personally interested in research and anxious to do all in their power to further the work, pay occasional visits to the various Institutes receiving State grants. The Development Commissioners publish annual reports dealing with the grants made and with the general nature of the investigations in progress. The feature of the British organization is the absence of direct Government control, even in the case of the Natural History Museum at South Kensington which is financed entirely by the State. In this case, the administration is in the hands of Trustees, who stand between the Government and the workers. This arrangement is said to be eminently satisfactory and to work well in practice.

In countries where local institutions do not exist, official Agricultural Departments have been created by the State. Such agencies organize the workers, distribute the funds, and arrange for the publication of results. The investigators are all Government officials and are generally entitled to a pension. The system possesses a number of advantages from the point of view of the State and of the workers. Government obtains credit for having devoted funds to the improvement of agriculture; the results obtained are a partial return for the revenue contributed by the farmers. The tax-payers feel that they are obtaining something tangible for their money beyond the maintenance of the ordinary amenities such as law and order, communications, and educational facilities. From the workers' point of view, a State system provides reasonable security of tenure as well as funds for the current work.

The direct control of research by the State is, however, attended by a number of disadvantages to the Government, to the public, and to the workers. When research is financed from annual grants, everything goes well as long as funds are plentiful. In periods of depression, however, there is always a cry for retrenchment which leads to the division of governmental activities into two classes—essential and non-essential. Research work, on these occasions, is apt to find itself in the second of these two categories and to be regarded as material for retrenchment. Alternations of expansion and retrenchment are apt to succeed each other with a frequency which is bewildering and which interferes with ordered progress. Another disadvantage from the financial point of view is the high cost of administration. Both men and money are absorbed by the administrative machine for which there is no direct return in the form of results. The great difficulty, however, when research is undertaken by the State is to accommodate the investigators in the official system. The origin of this difficulty will be understood, if the work of a Government office and that of an investigator are compared. The work of a State Department is very largely of a routine character and has to be carried out in conformity with the policy of the Government of the day, with various Acts of the Legislature and with the rules and codes based thereon. These activities are regulated by the budget, which is necessarily rigid and which has to be framed more than six months in advance of the official year. In addition, there is always a good deal of control by that section of the Government which deals with money. It is obvious that in such work the system must always be far more important than the individual and that there is very little room for new ideas. Anything published by a Government Department is binding on that Department at least, and often affects other branches of the Administration as well. There is, therefore, no room for individual opinion in official pronouncements, and it naturally follows that, in official publications, it becomes a safe rule to play for safety and to take the middle course. Very different is the work of a scientific investigator. Here the man is everything, the system is nothing. Official views do not exist. At any moment in research, a discovery may be made by the last joined recruit which amounts to a revolution in ideas, in outlook, and in methods of investigation. Thus the work of a Government Department, the men who carry it on, and the system under which it is conducted have little in common with scientific research, with scientific investigators, and the conditions which obtain in the great centres of creative work such as the universities. Both systems are useful in their own way, but it is sometimes difficult to combine them. Under a State system, the public interested in such subjects as agricultural research tend to become critics rather than helpers, and their energies are then apt to run into unprofitable channels.

In countries where local institutions do not yet exist, is any modification of the State control of research possible? Can the needs of the workers and the interests of the public be reconciled, and can some system be devised which approaches that of the best universities? It is suggested that a possible solution would be to separate all research, now carried on by the State, from the ordinary machinery of Government. A Development Board and a Development Fund could be created; this body could carry on the supervision of research now done by Government. The members of the Development Board could be elected partly by the Legislature, partly by the Executive, and partly by the workers themselves. The members of the Board would act as Trustees in the work of nation-building, and would occupy a position of great honour and responsibility. The Development Fund would have to be built up gradually from surplus revenue and, at the beginning, would require annual grants in addition. Under such an organization it would be possible to safeguard the workers and the public and to further the growth of local institutions. Many private individuals, anxious to assist in development work, could be readily incorporated in such a system and enlisted as helpers rather than critics of the work.

It would be interesting if such an experiment could be tried, and if the results obtained could be compared with those under a State system pure and simple. Developments in this direction, in regard to cotton growing, are being worked out at the present time. The research work on cotton, needed in many parts of the Empire, is being fostered by the Empire Cotton-Growing Corporation. Similar work in India is being done under the direction of the Central Cotton Committee. Both derive their income from a tax on raw cotton, and in both the State, the trade, the growers, and the scientific workers are directly represented.

Whatever the form of organization adopted, the method of recruitment and the working conditions must be such that the very best talent the nation possesses is attracted and retained. The development of a country's agricultural resources is of fundamental importance to the well-being of its people. In the past, it has often been assumed that men, insufficiently qualified for research in pure science, are quite suitable for applied work. The magnitude and complexity of the problems to be solved, however, are rapidly removing this impression, and it is now being realized that the very best products of the universities are needed for applied work. The attractions offered must, therefore, be such as to secure the best brains.

What are the conditions demanded by the successful scientific investigator? Such workers require freedom in the conduct of research and in the publication of their ideas; reasonable security of tenure on account of the protracted nature of the work; sufficient money to ensure a care-free life, and some method of reward or of recognition for success. The first three of these requirements—freedom, security of tenure, and reasonable remuneration—are gradually being obtained all over the world, but the last has not yet been satisfied. Indeed, a difference of opinion still exists as to the necessity and to the practicability of rewarding work of outstanding merit. The probability, however, is that, unless this defect is remedied, sufficient men of ability will not undertake research on agricultural problems. Without such men it is a question whether it is worth while taking up the work at all. Experience shows that professions like law and medicine, which offer outstanding possibilities of advancement and remuneration, and pure science, which is rewarded by academic distinctions, attract the best talent in the universities. Agricultural research has to be content with a few enthusiasts. From the nature of the work, these men have little time for the kind of research which brings academic recognition. While they have to sacrifice the rewards open to workers in pure science, they often make discoveries from which the country reaps benefits which may run into large sums of money. If some method of reward could be devised for applied work of this character, a great step forward would be made. That country which is able to solve this question successfully is certain to attract some of its best talent for the development of its agriculture. It is often contended that love of work and joy in discovery are sufficient reward for the investigator of such questions, and that nothing further is required. This method, however, has not been successful in the past, and sufficient workers of outstanding ability have not taken up the work. At present the Empire is exceedingly short of workers capable of dealing with crop questions. On all sides problems awaiting investigation are lying ready to hand, and in many cases funds are available for the work. Appointments abound, but there are not enough qualified men for the work. One reason would seem to be that the rewards now offered for success are not sufficient to attract the best talent in the universities.

## BIBLIOGRAPHY

Report of the Development Commission for the year ended the 31st March, 1923, H.M. Stationery Office, London, 1923.

Leake, H. M.—The organization of agricultural research in India, *Agr. Jour. of India*, xiii, 1918, p. 583.

Perkins, A. J.—Some suggestions as to the Commonwealth Endowment of Agricultural Research, *Bull. 7, Advisory Council of Science and Industry*, Melbourne, 1918.

Report of the Indian Industrial Commission, Simla, 1918.

Report on Scientific Property, League of Nations, Geneva, 1923.

# INDEX